Clearing the Air: Federal Policy on Automotive Emissions Control

Clearing the Air

Federal Policy on Automotive
Emissions Control

Henry D. Jacoby,
John D. Steinbruner,
and others

Ballinger Publishing Company • Cambridge, Mass.
A Subsidiary of J.B. Lippincott Company

Copyright © 1973 by Ballinger Publishing Company. All rights reserved. No part of this publication may be reproduced, stored in a retrieval system, or transmitted in any form or by any means, electronic mechanical, photocopy, recording or otherwise, without the prior written consent of the publisher.

Library of Congress Catalog Card Number: 73-10393

International Standard Book Number: 0-88410-301-3

Printed in the United States of America

Library of Congress Cataloging in Publication Data

Main entry under title:
        Federal policy on automotive emission control.
        Includes bibliographical references.
        1. Automobile exhaust gas. 2. Air—Pollution—Law and Legislation—United States. I. Jacoby, Henry D. II. Steinbruner, John D.
TD886.5.F4   614.7'12   73-10393
ISBN 0-88410-301-3

Contents

Preface

Chapter One
Policy Analysis in the Public Sector   1
Henry D. Jacoby and John D. Steinbruner

The Dimensions of the Problem   3
The Structure of the Clean Air Act   4
Preview of the Analysis To Follow   6

Chapter Two
The Context of Current Policy Discussions   9
Henry D. Jacoby and John D. Steinbruner

History of Legislation   9
The Current Technical Situation   15
The Issue of Extending the Standards   25

Chapter Three
Policy Options and Predicted Outcomes   27
Henry D. Jacoby and John D. Steinbruner

Established Policy   29
Relaxed Standards   37
Alternative Technology   38
Evaluation of the Options   40

Chapter Four
Advanced Technology and the Problem of Implementation   49
Henry D. Jacoby and John D. Steinbruner

Historical Perspective 50
A Strategy for Implementation 54

Chapter Five
Emissions Measurement and the Testing of New Vehicles 63
Milton C. Weinstein and Ian D. Clark

Framework for the EPA Decision-Making on Testing 64
Structure of the Analysis 66
Measures and Measurement: What to Measure and How to Do It 68
Concepts Used in the Analysis 83
Prototype Testing 92
Assembly-Line Testing 95
On-The-Road Testing, Recall, and Warranty 98
Appendix: Statistical Models Used in the Analysis 105

Chapter Six
Testing and Maintenance of In-Use Vehicles 113
Jack M. Appleman

State Implementation of Enforcement Programs 114
Manufacturers' Response to Warranty and Recall 120
A Model of Auto Emissions Inspection and Enforcement 123
Conclusions 137

Chapter Seven
Health Effects of Automotive Air Pollution 139
William R. Ahern, Jr.

Background on Air Pollution 141
Exposure 147
Background on Health Effects of Pollutants 148
The Analysis Behind Current Emissions Standards 162

Chapter Eight
Measuring the Value of Emissions Reductions 175
William R. Ahern, Jr.

An Approach for Valuing Reductions in Auto Emissions 177
Research on Health Effects of Pollution 203
Conclusion 205

References 207
Index 211
About the Authors 215

List of Tables and Figures

Figure

3-1	Schematic View of Alternative Policy Choices and of Uncertain Outcomes Under Each of the Policies	28
3-2	Emissions Rates for Different Stability Levels on the Assumption That No Specific Emissions Control Maintenance Is Performed, Stated As a Multiple of the Prototype Test Result	31
3-3	Nationwide Carbon Monoxide Emissions for Alternative Vehicle Stability Levels Under Established Policy, With No Program of Inspection and Enforced Maintenance	32
3-4	Expected Outcomes Under Alternative Policy Options for Reasonable Assumptions About Program Cost and Stability of Vehicle Emissions	44
3-5	Expected Outcomes Under Alternative Policy Options for Reasonable Assumptions About Program Cost and Extremely Optimistic Assumptions About Vehicle Stability	46
5-1	Framework for Decision-Making on Testing	65
5-2	Variation in Observed Test Result	84
5-3	Likely Manufacturer Response to Test Variation	85
5-4	Adjustment of Nominal Standard To Achieve Target Level	87
5-5	Typical Patterns of Emissions Deterioration	91
6-1	Model of Vehicle Emissions As a Function of Miles Driven and Number of Years Without Specialized Service of Emissions Controls	127
7-1	Descriptive Model of the Process Relating Health Damage To Auto Emissions	140
7-2	Atmospheric NO_2 Photolytic Cycle	143

7-3	Interaction of Hydrocarbons with Atmospheric NO_2 Photolytic Cycle	144
7-4	Model of a Commuter's Exposure to Air Pollution	148
7-5	The Main Anatomical Features of the Respiratory System (Left) and the Terminal Bronchial and Alveolar Structure of the Lung (Right)	149
7-6	Schematic Portrayal of Pollutant Deposition Sites and Clearance Processes	151
7-7	Descriptive Model of the Barth Analysis, Taking CO as an Example (CO Distribution from Camp Data)	165
7-8	Maximum Daily One-Hour Average Oxidants as a Function of 6:00 to 9:00 a.m. Averages of Nonmethane Hydrocarbons at CAMP Stations, June Through September, 1966 Through 1968, in Philadelphia, Denver, and Washington, D.C., and at Los Angeles, May Through October 1967	169
7-9	Maximum One-Hour Average Oxidant Concentrations as a Function of 6:00 to 9:00 a.m. Averages of Total Nitrogen Oxides in Washington, D.C., June Through September, 1966 Through 1968, and in Philadelphia and Denver, June Through September, 1965 Through 1968	170
7-10	Approximate Isopleths for Selected Upper-Limit Maximum Daily One-Hour Average Oxidant Concentrations as a Function of the 6:00 to 9:00 a.m. Averages of Nonmethane Hydrocarbons and Total Nitrogen Oxides in Philadelphia, Washington, D.C., and Denver, June Through August, 1966 Through 1968	171
8-1	Outline of an Approach for Valuing Reductions in Auto Emissions	178
8-2	Eight-Hour Averaging Time Carbon Monoxide Concentrations (ppm) Exceeded 0.1 Percent of the Time in the Los Angeles Area, 1956 Through 1967	180
8-3	Effect of Reductions in 1967 CO Emissions on CO Levels for a Representative Central City	182
8-4	Effect of 50 Percent and 75 Percent Reduction of CO Emissions per Car on CO Levels for a Representative Central City in 1990	183
8-5	1967 Oxidant Concentrations and the Effects, by 1990, of 50 Percent and 75 Percent Reductions in Oxidant Precursor Emissions (HC and/or NO_x) for a Representative Central City and One-Third of the Suburban Area	186
8-6	Oxidant Transport over a Representative Urban Area	187

8-7	Graphs of Total Annual EDRA's Associated with Various Reductions in CO and Oxidant Precursor Emissions	199
8-8	Marginal Annual EDRA's Attributed to 1 Percent Reductions in 1967 CO and Oxidant Precursor Emissions	200

Table

2-1	Estimates of the Increase in First Cost and Fuel Consumption over 1973 Vehicles to Meet 1975 and 1976 Emissions Standards	19
3-1	Weighted Index of Reduction in Emissions Under Different Policy Options, for Indicated Stability Levels for ICE Vehicles and Level 0 for Alternate Technology	33
3-2	Low, Medium, and High Estimates of Fifteen-Year Program Costs under Established Policy at a 5 Percent Discount Rate	36
3-3	Low, Medium, and High Estimates of Fifteen-Year Program Costs Under Relaxed Standards at a 5 Percent Discount Rate	37
3-4	Low, Medium, and High Estimates of Fifteen-Year Program Costs with a Shift to Alternative Technology in 1981 at a 5 Percent Discount Rate	40
3-5	Expected Outcomes of Alternative Policy Options Under Reasonable Assumptions About the Relative Likelihood of Different Levels of Cost and Vehicle Stability	42
5-1	Estimated Net Change in De Facto Standards Produced by Hot Start-Cold Start Rule Change	79
6-1	Definition of Alternative Stability Levels	129
6-2	Weighted Index of Reduction Under Established Policy Under Alternative Weighting Schemes	132
6-3	Weighted Index of Reduction Under Alternative State Enforcement Schemes (w_k = .12, 1.0, 1.0)	136
7-1	Carbon Monoxide Concentration (in ppm) at CAMP Sites for an Eight-Hour Averaging Period, 1962-1967	167
8-1	Distribution of Carbon Monoxide Levels for a Representative CAMP Station in 1967 (Only Hours in Which the National Standard Is Exceeded Are Shown)	181
8-2	Cumulative Frequency Distribution of Hourly Average Oxidant Concentrations in Selected Cities, 1964-1965	185
8-3	Distribution of Oxidant Levels for a Representative CAMP Station in 1967 (Only Hours During Which the National Standard Is Exceeded Are Included)	186

8-4	Sample Factors for Converting Health Effects to EDRA's	191
8-5	Estimation of the Number of Equivalent Days of Restricted Activity Attributable to Oxidant Concentrations at 1967 Levels and for 50 Percent and 75 Percent Reductions from Those Levels (Susceptibility Category: Persons (Cases) with Chronic Conditions, Age 0-64, Central City Exposure)	192
8-6	Estimation of the Number of Days of Restricted Activity Attributable to Carbon Monoxide Levels for 1967 Levels and for 50 Percent Reductions: (Susceptibility Category: Persons (Cases) with Acute Conditions, Central City Exposure	193
8-7	Total National EDRA's Assigned by Three Estimators to Chronic and Acute Health Conditions for Three Pollutant Concentration Profiles of CO and Oxidants	194
8-8	High Annual National Health Damage Estimates Using Eight-Hour Averaging Period When CO Concentration Exceeds 10 mgm/m^3	195
8-9	Low Annual National Health Damage Estimates Using Eight-Hour Averaging When the CO Concentration Is 15.5 mgm/m^3 or Higher (Figures Given for the Central City Only)	195
8-10	High National Oxidant Damage Estimates	197
8-11	Low National Oxidant Damage Estimates	197
8-12	Total High and Low EDRA Estimates for CO and Oxidants, in Millions of EDRA's Per Year	198
8-13	High and Low Valuations of High and Low National Annual EDRA Estimates Attributed to CO and Oxidants, in $ Millions	203
8-14	Incremental Values of National Annual EDRA's Prevented by Reducing 1967 Emissions by 50, 75, and 90 Percent, in $ Millions	204

Preface

This study was undertaken as a joint effort of two research projects at Harvard University. The Environmental Systems Program, under sponsorship of the National Science Foundation, is carrying out research on methods of analysis of a broad range of environmental problems. The Institute of Politics of the John F. Kennedy School of Government—through its Faculty Seminar on Bureaucracy, Politics and Policy—is working toward a better understanding of the way bureaucratic institutions affect policy outcomes and toward the development of a better set of methods of analyzing problems of policy implementation in the public setting. The Institute Seminar has been supported by the Ford Foundation.

In the summer of 1970 it was decided that the two research groups would take up the automotive pollution issue as a joint venture. The Environmental Systems Program is an interdepartmental activity with participation by faculty and students from the Division of Engineering and Applied Physics, the Kennedy School, the Department of Economics, and the Graduate Schools of Design and Public Health. The Program traditionally has placed a heavy emphasis on environmental engineering, systems analysis, and economics. The Seminar on Bureaucracy, Politics and Policy has its strength in the areas of organizational behavior, policy studies, and law. The auto air pollution issue was chosen precisely because it involves a rich mixture of science, technology, and economics on the one hand and of politics and policy on the other. Shortly after work began on the topic the 1970 Amendments to the Clean Air Act was passed, creating an excellent opportunity for the type of policy study we were seeking to undertake.

We have pursued a three-fold objective in this study. In the first instance we have followed the auto air pollution issue as closely as possible, and have tried to provide analysis that will be useful to persons involved in making public choices in this area. With such a complex issue, this is a difficult task in the best of circumstances, and in this regard a university setting has both advantages and disadvantages. On the one hand, it has been possible to step back and take a long-term, broad view of the issue (a luxury often forbidden analysts on the firing line in public agencies). On the other hand, we do not have the authority or the information resources of the responsible organizations. No doubt we suffer from limited personal involvement in the rapid day-to-day development of these issues in the committees of Congress, the EPA, the auto industry, and in the courts. In seeking to capture the advantages of academic detachment, we have attempted to bring as much reason as possible to the issue without intentionally becoming partisan for or against any of the parties to the heated public debate.

Apart from the value of the final product, there was a second reason to undertake a relevant piece of policy analysis within the time deadlines imposed by the unfolding events themselves, which is to learn how to do a better job of this kind of analysis of public choices. We have used the auto air pollution issue as a laboratory where we could observe ourselves grappling with problems of analysis of policy formation and implementation. Certainly we have learned from our successes and failures. Some day there may be a neatly codified body of theory and method in the field of policy analysis. At this point, however, we are still at the stage where the best way to develop new knowledge of the tools and techniques of policy analysis is to go out and do it in as true a simulation of real-world conditions as one can manage. Therefore, this book is really an extended case study of a complex public policy issue. It is our hope that the study can serve as case material for the teaching of policy analysis, to be used or criticized as appropriate.

Finally, one of the important purposes of this type of research activity in a university setting is to create opportunities for joint faculty-student work and thereby to train students. It is important to emphasize the extent of student contribution to this study and to acknowledge some of the many people who participated in it.

First, in the summer and fall of 1971, when we were first digging deeply into the issue, critical input was made by Carol Kellerman and Andrew Spindler—Harvard students who were residents in Washington, D.C., in the summer of 1971 and who conducted interviews and gathered materials that would have been inaccessible without having people on the scene. We wish to thank the Center for the Study of Responsive Law, which provided office space to Kellerman and Spindler for the summer. Susan Lutzker, a former research assistant at the Institute of Politics, also covered events in Washington, D.C., and in New Jersey during the summer of 1971; and early in the study we benefited from the assistance of T.A. Moen, a student in the Division of Engineering and Applied Physics, and Edward Delaney, a student at the Harvard Law School.

During the months when we were first familiarizing ourselves with the issue, invaluable input was made by Stephen Moeller, first as a student in the Harvard Law School and the Kennedy School of Government and then as a research assistant. His skill in gathering data and in sifting and sorting out technical issues provided much of the gist for chapters that ended up being written by others. At a point when the study was more advanced, Mark Lowenthal, a graduate student in history at Harvard, prepared a paper on the conversion of the automobile industry to war production that deepened our understanding of the industry.

Since both of us were associated with the Public Policy Program of the John F. Kennedy School of Government, the auto air issue was used as workshop material in the fall and winter of the 1971-72 academic year. In January of 1972, automotive pollution was used as the subject for an exercise involving all the first-year Public Policy students. It was an exercise that greatly broadened our experience with the issue. Also various aspects of the problem were taken up by second-year students in the Program. The work by William Ahern, Jack Appleman, Ian Clark, and Milton Weinstein ultimately has become part of this book. Contributions by Don Howarth (on antitrust aspects), Maureen Steinbruner (on research and development programs), Christian Stoffaes (on gaseous fuels), and Leon Loeb (on electric drive vehicles) formed part of the general background of information which stands behind an effort like this.

In addition to these students, there are several members of the Harvard staff who deserve a special thanks for help at critical points in the study. In particular, we are indebted to Tom Shemo for assistance with computer programming, Nancy L. Reynolds and Eileen Kantrov for editorial assistance, Renée Chotiner for early research assistance, and most especially to Susan Ackerman, who has been an energetic, effective research assistant and project expediter throughout the work.

An earlier product of this study was an article entitled "Salvaging the Federal Attempt to Control Auto Pollution" published in the Winter, 1973 issue of *Public Policy*, which served as the nucleus for Chapters 1 through 4 of this book. We thank *Public Policy* for permission to use this material in the current work.

Finally, we must acknowledge the many persons who gave advice and help along the way or who gave a critical reading of early drafts. Special thanks are due to Howard Raiffa, Philip Heymann, Robert Dorfman, Stephen Moeller, and Helen Kahn, and to the staff members at the Environmental Protection Agency, General Motors, and Ford who gave help at the start and helpful criticism at the end. Naturally, the fact that we benefited from this help does not necessarily imply that they will agree with each step in our analysis or with our conclusions.

Henry D. Jacoby
John D. Steinbruner
Cambridge

Clearing the Air: Federal Policy on Automotive Emissions Control

Chapter One

Policy Analysis in the Public Sector

Henry D. Jacoby
and
John D. Steinbruner

In the 1970 amendments to the Clean Air Act[1] the United States has embarked on a major adventure in government regulation. The amendments mandate drastic reductions in the air pollutants emitted by automobiles and require that these reductions be achieved by the 1975 and 1976 model years. These regulations strike at one of American society's fundamental technologies and impose strict discipline on its largest industry. More than 10 percent of the United States Gross National Product is generated by the production, fueling, and servicing of the internal combustion engine (ICE in the technical vernacular) in its automotive application; the associated social and political forces are commensurately large. The enactment of this legislation over the opposition of the automobile industry was an impressive display of public resolve, and the fact that the industry is now actively working to comply is a major victory for the public interest.

Unfortunately, the regulatory mechanisms set up in the Clean Air Act are too primitive for the complex technical and manufacturing processes to which they have been applied. Unless adjustments are made the results are likely to be unhappy. Under the current policy, consumers and taxpayers will pay an unnecessarily high price for the emission reductions that are actually achieved, and even then the objectives set in the legislation will not be met.

The underlying difficulty has to do with the relationships among the emissions standards, the short deadlines, and the state of the art of automotive technology. Controlling the conventional ICE to meet the new requirements will involve sharply increased manufacturing and operating costs and noticeably reduced performance on the road. In order to avoid these severe trade-offs it would be necessary either to introduce an entirely different automotive technology or to make fundamental changes in the design of the internal combustion

---

[1] Title II of the Clean Air Act as amended December 31, 1970, contains the regulation of automobile pollution. That Title is officially labeled the National Emissions Standards Act.

engine. Promising alternatives are known to the scientific community, but the manufacturers—who are financially, technically, and even psychologically committed to conventional engines—have focused most of their technical efforts on the standard ICE. They have been reinforced in this behavior by the 1975 and 1976 deadlines, since it is unlikely that any of the alternative technologies could be prepared for mass production by that time under constraints on cost and effort generally accepted as legitimate. The result is an abatement program that is based almost exclusively on marginal adjustments and bolted-on additions to the conventional ICE.

With the pursuit of pollution control effectively constrained within a relatively narrow and unpromising technical area, four serious problems arise:

1. Approaching the 1975-76 standards with a controlled ICE is likely to be expensive, not only in initial manufacturing cost but also in terms of fuel and maintenance expenditures.

2. Despite this expense there is a good chance that the conventional ICE cannot meet the 1975-76 standards, even by the late 1970s.

3. Even if the standards are met under the current official definition (which involves emissions tests on prototype vehicles), emissions rates on the road are likely to be highly unstable. That is, the average emissions of the actual vehicle population can be expected to be significantly higher than the average emissions rate calculated on the basis of official test procedures.

4. Control of a highly unstable technology, once it is in the hands of users, would require elaborate federal and state enforcement machinery. Such an effort might involve governmental surveillance and regulation of tens of millions of individual vehicles and tens of thousands of firms in the service industry. Such systems would be expensive to set up, and their value in actually contributing to cleaner air is highly questionable.

Moreover, under the established policy the natural course of events will run to the relative advantage of the automobile manufacturers. The cost of producing automobiles will increase by a significant amount, but the industry has sufficient financial flexibility to avoid disruption of the market for new cars. The major costs of emissions control promise to arise from increased fuel consumption and maintenance—costs that fall on the consumer after he has purchased the car. Any consumer wrath at these costs is as likely to be directed against the government as against industry. Further, in the event that the manufacturers fail to meet the standards, it is the government that will have to yield. The demonstrable benefits to be gained from limiting automobile emissions to the low levels required are not great enough to justify the immense economic, social, and political costs of a major disruption in automotive production. No administration could withstand the political pressure that would result.

In the following pages we will be concerned with how the situation came to such an impasse and with an analysis of what might be done about it. It should be emphasized, however, that there is more to the exercise than a dissection of the immediate issues, important though they are. The automobile air pollution issue represents a larger class of problems that seem to plague modern society: it requires the regulation of very large organizations; it involves a conflict between public and private interests; it entails a strong interaction between political, economic, and technical factors. In pursuing an analysis of the policy on automotive air pollution we seek also some improvement in the art of analysis with respect to the entire class of problems.

THE DIMENSIONS OF THE PROBLEM

The issue of automotive emissions control is one that the newspapers like to refer to as "complicated." It pays to ask just what that deceptively familiar word might mean, for the label covers a multitude of things. In the first place, automotive emissions are an example of what economists call "external effects"—that is, effects of economic significance that are not included in the payment arrangements of the marketplace. A person who wishes to drive an automobile must pay for its use, thereby providing direct compensation for the labor, capital, and materials consumed in its production and operation. On the road, however, he puts pollutants into the atmosphere which damage those about him, and there is no arrangement to make him pay for that. A related dimension of the problem is what is sometimes called the "crisis of the commons." The contribution of any one automobile to overall air pollution is exceedingly small, whereas the cost involved in eliminating his small contribution is quite noticeable to the individual motorist. It appears rational in terms of any person's own self-interest to avoid paying the cost and to depend on others to reduce pollution. Unfortunately, when everyone behaves in terms of such a calculus, a situation results in which no clean-up is accomplished, the common air is polluted, and everyone is damaged. In circumstances like this, private interests, rationally pursued, produce public disasters.

Moreover, in a complicated world nearly everything turns out to be related to everything else, and this is particularly true when one is dealing with a technology as important as the internal combustion engine. Taking an analogy from the game of pool, if one hits the cue ball, the other balls will ultimately move in an extensive series of collisions and bounces. The challenge of the game lies in the fact that it is difficult to limit the movements subsequent to those which are intended. The same is true, on a vast scale, in the business of intervening in automotive technology. There will be many more adjustments in the state of society than simply the reduction of automotive emissions, and not all of these will be desirable.

Finally, the presence of systemic effects intensifies two other important dimensions of the problem: uncertainty and value trade-offs. In pool it takes a great deal of experience to be able to understand what all of the consequences of hitting the cue ball in a particular direction at a particular speed are likely to be. By the second or third strike the likely movements are not intuitively obvious. When one expands the problem to encompass a society of 200 million people it becomes hopelessly complex, and hence there are many consequences of public actions—even major consequences—that are not understood in advance. Some of the surprises inevitably are unpleasant ones. Man as a limited being operates under great uncertainty, and so do all of his institutions. This fact is a major feature of the human condition and has a major impact on behavior.

The matter of trade-offs identifies the unhappy truth that actions taken in pursuit of one purpose in a highly interactive system regularly produce effects that directly or indirectly damage other purposes. First of all, scarce resources are consumed that might be devoted to other ends. More troublesome is the evidence that, depending on how it is done, the effort to control automobile emissions may interfere directly in otherwise separate areas of concern—in employment, for example, in the overall use of energy, and in the balance of payments.

THE STRUCTURE OF THE CLEAN AIR ACT

In the face of such wonderful complexity, the solution embodied in the Clean Air Act has appealing simplicity. The Congress has established the goal that no citizen's health should be injured by air pollutants spewed out by others. The automotive portion of the Act proceeds in a straightforward manner to establish the degree of emissions reduction needed to attain this result, and forbids the sale of vehicles that fail to meet the associated pollution allowances. This approach is part of a long tradition of regulatory policy, stretching from the earliest health standards for milk to the insulation requirements for electrical wiring. Very often the government need not involve itself in the process being regulated save to set the appropriate limits, specify when they must be achieved, and establish a procedure to enforce compliance.

It should be noted, by the way, that this approach to problems of externalities is not without a strong intellectual foundation. To use the language of economics, when an external effect presents a social problem, then market arrangements should be revised to incorporate or "internalize" them. There are several ways to do this: for example, by various forms of fees or fines, by subsidies, or (as in the case of most environmental problems) by imposing direct limits on private behavior. Whichever scheme is used, the market mechanism usually is depended upon to solve the many complex problems outlined earlier. If markets work smoothly, then the level of production of the external effects will find an equilibrium level with other values reflected in the economy, and the multitude of unforeseen and unforeseeable consequences will be worked out automatically in price adjustments. Competition within the market and the natural selection associated with it ultimately will assure that the most efficient treatments of the external effects are adopted.

This view of the logic behind the Act can be seen clearly in the position taken by congressional staff members who had been deeply involved in drafting the current auto pollution control legislation. In the early months after the 1970 amendments were passed, they argued that it would be a waste of the taxpayers' money to pursue more federal emissions control research. The manufacturers, in their view, would have to comply or they could not sell cars; in the interest of profits they could be expected to choose the most efficient way to meet the standards. Thanks to the Act, the development of clean technology was no longer seen as the government's problem; the responsibility, they felt, had been passed to the industry.

We shall argue that this was an unfortunate viewpoint in the peculiar circumstances of the automotive case. The simple regulatory approach does allow the government to remain free of many of the details of a complicated problem, and it fits well with American traditions and habits of mind. But if the regulatory device and actual market conditions fall drastically short of the theoretical ideal, then there are grounds for worrying whether the policy will in fact produce the desired result.

In the automotive case, for example, one cannot help but notice that we are not dealing with a multitude of independent firms as in the conventional picture of the free market mechanism. Emissions regulation is imposed on the automobile manufacturers, who are few in number and who behave not remotely like atomistic individuals but rather like what they are—very large organizations. Competition within the industry has a very special character that can produce unexpected reactions to a severe market constraint such as the 1975-76 emissions standards. The behavior of these manufacturers cannot be satisfactorily predicted simply by assuming that they will strive to maximize profit under the newly supplemented market arrangements. These organizations operate on the basis of long-established procedures and mental sets. Powerful inertial forces will not necessarily be channeled in the intended direction by simple emissions standards and deadlines. Finally, the companies have great potential to exercise political power and hence could produce a wide range of possible effects in that sphere. (These same arguments apply as well to state governments that are to assist in enforcement of the standards on the road.) There is, in short, a great deal that could go wrong with the current policy. It is vulnerable to forces whose importance is masked by formulations of the problem based on simple economics and no more.

These doubts all culminate in a two-part argument. First, the government should not count solely on a procedural rule such as the current emissions standards and deadlines to produce a happy outcome. And second, by contrast, the government should decide directly upon the desirable *outcomes* and should take steps to achieve them directly. There should, in other words, be a shift in the basis of policy from reliance on procedure to direct outcome calculation. The distinction is a subtle one, for the two categories are not mutually exclusive. But it is of fundamental importance, for it alters the logic upon which the policy rests, generates new questions for government analysts to answer, and imposes far more demanding standards of analytical performance.

A policy relying on direct outcome calculations, for example, would have to consider in detail the alternative consequences of the emissions regulation—tracing out what control techniques will in fact be used and exploring a range of possible emissions results and cost profiles. The intentions of policy designers are rarely if ever carried out exactly, and never do these intentions define the only possible outcomes. A policy based on direct outcome calculation is inevitably driven to consider the range of unintended and undesirable outcomes and to hedge against them. Such a policy stance is also likely to use a variety of mechanisms for influencing the course of events, to be driven to coordinate the workings of these mechanisms, and to engage in reasonably sophisticated monitoring of what is actually happening. In general, direct outcome calculation requires a far more detailed treatment of substantive issues than does the application of standard procedural rules, and a far more elaborated, more complete policy.

This distinction between a policy based on direct outcome calculation and one based on procedures is a major theme in the study that follows, but it is inevitably vague and confusing when stated abstractly. Hence the focus in the subsequent chapters is on the particular problem of automotive emissions control. This problem provides a more concrete and, at this point, more useful context in which to pursue the general issue.

PREVIEW OF THE ANALYSIS TO FOLLOW

To lay the groundwork for our analysis, Chapter 2 reviews the history of the federal efforts to control automotive emissions and surveys the existing knowledge about the technical issues that loom so large in this case. In Chapter 3, various policy options are developed and an attempt is made to evaluate each in terms of the outcomes they are likely to produce, taking into account the current uncertainty about the costs of different control systems and their likely performance over time.

The analysis in Chapter 3 shows that the current policy, when evaluated in terms of ultimate outcomes, is clearly dominated by other approaches involving a greater public involvement in the research on a clean alternative engine and in the detailed steps leading to its adoption. Chapter 4 then explores some of the barriers to speedy adoption of advanced engine designs and plots the details of a plan for implementing the more desirable emissions control policy. Our conclusion on this score is that the current Act should be amended to provide for more direct federal involvement in research on alternative propulsion systems. We also propose a new set of financial incentives intended to spur adoption of the control systems that will result from existing and future research.

Chapters 5 and 6 consider many of the details that must be taken into account if an analysis of automotive control policy is to look all the way down the path of events to the expected outcome. One important aspect in this case, naturally, is the pollution emitted by cars, and therefore a key element in the implementation of control policy is the emissions measurement and testing system that *defines* what this outcome is. Even the definition of emissions is far from a simple matter, and Chapter 5 begins with a hard look at the many problems of instrumentation and aggregation that lie behind a statement that a car emits a particular number of grams per mile of some pollutant. Chapter 5 goes on to construct a logically coherent picture of the testing and enforcement options that are available to the government at the three stages in a vehicle's life—in prototype, at the assembly line, and on the road. Prototype testing is required for certification of new vehicle designs. If control systems are so fragile that cars rolling off the assembly line may not resemble the cars presented for certification, then systems of assembly-line monitoring must be developed. If emissions control systems are prone to deterioration under the rigors of road use, as is the conventional ICE, then systems of on-the-road testing and enforced maintenance will be needed in order to meet the strict emissions targets. Testing at all three of these points is provided for in the Clean Air Act, and they must be coordinated if the desired level of performance is to be achieved. Chapter 5 makes a somewhat optimistic set of assumptions about the rationality of manufacturers' behavior and about the capacity of state governments to carry out inspection schemes and shows the necessary conditions for a coherent set of enforcement actions.

Chapter 6 looks deeper into the on-the-road enforcement systems that the states might establish. It takes a more critical view of the willingness and ability of the states to implement in-use testing schemes and of the capacity of the service industry to perform the repairs that may be called for. The results further support the conclusions reached in Chapter 3 regarding the need to change policy in ways that encourage the adoption of alternative engine technology.

Chapters 6 and 7 take up a set of outcome calculations in the dimension of health effects. As noted earlier, the whole control policy is based on a decision that adverse health effects are to be avoided. The critical importance of this aspect naturally raises the question of how one might go about determining what the effects are. Again, this is far from a simple matter, and Chapter 7 is devoted to a review and analysis of the current state of knowledge on this issue. Chapter 8 then develops an analytical structure for estimating the benefits to health of marginal changes in emissions standards and presents some preliminary estimates. It is the kind of analysis that is needed if a policy is to be judged on the basis of outcomes (i.e., health effects achieved) rather than on the apparent toughness of procedural rules.

So, throughout the book the orientation is the same. Automobile emissions control is a very messy situation. In the formation of policy in this area many things can go wrong and are going wrong right now. No doubt some of the confusion and waste is unavoidable, but in this case a more direct government involvement in manipulating the outcome, supported by analysis of the likely results of different actions, could help avoid some of the more gross inefficiencies. This book is offered as a modest attempt to do policy analysis at that level of detail.

Chapter Two

The Context of Current Policy Discussions

Henry D. Jacoby
and
John D. Steinbruner

A tough, tightly specified federal policy on the control of atmospheric pollutants from the automobile—emissions, in the technical jargon— was established by amendments to the Clean Air Act which were signed into law on December 31, 1970. That legislation, far more drastic than most informed observers had expected, was the product of at least twenty years of skirmishing between government and industry. To understand the legislation it is necessary to understand its historical development.

HISTORY OF LEGISLATION

The issue arose in 1950, when studies at the California Institute of Technology first clearly associated automobile emissions with smog in the Los Angeles basin. The researchers discovered that photochemical reactions involving hydrocarbons and oxides of nitrogen in the presence of sunlight produce a number of compounds which are irritating and potentially dangerous to sensitive human tissue, particularly lungs and eyes. Shortly thereafter, the Los Angeles Air Pollution Control District began encouraging the automobile companies and the state government to take corrective action.

In these early days, the manufacturers repeatedly denied that automotive emissions were a problem. Nevertheless, they did set up a study group within the Automobile Manufacturers Association (AMA) to conduct research and exchange technical information. And in 1954 they reached a cross-licensing agreement among themselves which provided for general use, without royalties, of any air pollution control equipment developed by any of the auto manufacturers.

The first federal legislation to control air pollution came in 1955 with PL 84-159 which located responsibility in the Department of Health, Education, and Welfare (HEW). This law provided for federally sponsored research but pointedly did not grant the federal government any enforcement powers. The Public Health Service, which took over administration of the program, opposed federal enforcement powers and insisted that the responsibility resided with state governments. In 1959 bills were submitted to Congress prohibiting the use in interstate commerce of any vehicle emitting substances in amounts dangerous to human health, but actual legislation in 1959 only provided for a report by the Surgeon General on the effects of automotive exhaust.

In 1961 California passed a law requiring a simple crankcase control device and followed in 1963 with legislation requiring exhaust control devices on vehicles marketed in California as soon as two devices were approved by the State Motor Vehicle Control Board. The automobile manufacturers at first denied that they had any such control technology available. But when four devices submitted by independent manufacturers were approved by the California Board in 1964, the companies announced that devices of their own manufacture would appear on new cars sold in California beginning with the 1966 model year. The first California emissions standards were then imposed for that model year.

The earliest federal enforcement powers, which were established by the Clean Air Act in December of 1963 (PL 88-206), authorized the federal government to convene interstate pollution abatement conferences and in some cases to initiate federal suits to force reduction of emissions. It also provided for federal grants to stimulate research and to increase state pollution control activity, and it specifically mentioned the need for further attention to automotive exhaust. A second title to the Act was passed in 1965 authorizing HEW to set emissions standards for automobiles beginning with the 1968 model year. A year later HEW announced 1968 standards roughly the same as those that had come into effect in California in 1966. The impression was clearly conveyed that the federal government was assuming enforcement powers slowly and reluctantly, and that federal activity was lagging behind the more aggressive California program.

The political climate that ultimately precipitated the stringent emissions standards began to develop in 1965 when Ralph Nader published his famous indictment of the industry for safety hazards and was treated to a personal investigation at the industry's expense. Seldom has an attempt at intimidation backfired so spectacularly. The Nader affair led to a dramatic set of hearings in which the president of General Motors was forced to apologize to Nader in front of a congressional committee and a national television audience. Serious and lingering damage was done to the political credibility of the automobile manufacturers—damage soon compounded by allegations concerning their handling of the air pollution problem itself. In January 1965, the Los Angeles County Board of Supervisors requested that the Attorney General investigate collusion by the industry to withhold pollution control equipment. The supervisors charged that the committee of the AMA set up to conduct joint research was in fact a collusive arrangement to prevent the introduction of controls.

As evidence that industry developments were being suppressed rather than propagated, they cited the package of control devices developed by Chrysler but kept off the market until California legislation forced its introduction. The resulting Justice Department investigation ended in a consent decree in 1969 that provided for an end to the conspiracy without officially conceding its existence. This incident unquestionably added to the public's impression of recalcitrance and bad faith on the part of the industry.

In 1970 the political forces supporting stronger emissions controls finally became overwhelming. The press discovered the pollution issue and began to raise public consciousness. NBC News showed filmclips of diseased trees over 100 miles from Los Angeles. The *New York Times* added a special reporter for environmental affairs, and he and others like him soon announced that environmental protection was the coming issue on American campuses. Earth Day was proclaimed on April 22, and well-publicized nationwide activities called for stronger pollution control measures. Ralph Nader's Center for the Study of Responsive Law published a well-timed report on air pollution that attacked both the automobile industry and Senator Edmund Muskie, who was chairman of the Senate Subcommittee on Air and Water Pollution.[1]

The attack by Nader's study group stung Muskie, then a Presidential hopeful. It threatened him in an area where he had established a strong public reputation and caught him at a time when he was being pressured from several sides. President Nixon had sent a message to Congress in February calling for a 90 percent reduction in emissions standards by 1980 and proposing federally sponsored research on low-polluting automotive technologies. Whatever the President's actual intentions were, it looked very much like a deft political finesse on a major developing issue. Moreover, Senators Gaylord Nelson and Henry Jackson were issuing proposals for environmental protection which threatened Muskie's jurisdiction over the issue. In August, Nelson introduced a bill banning the internal combustion engine outright by 1975. Later that month, during a period of unusually high air pollution levels on the East Coast and with a transcontinental Clean Car Race in progress, Muskie took Nixon's standards and Nelson's deadline and fashioned his own program.

As chairman of the critical subcommittee, Senator Muskie was able to exert his authority over the legislation, and his initiative forced the Nixon Administration into a defensive position. In November 1970, HEW Secretary Elliot Richardson wrote a letter to Congress requesting that the deadlines in the Senate bill be relaxed. William Ruckelshaus, nominated as head of the newly formed Environmental Protection Agency (EPA), asked for authority to extend the deadlines; and Edward David, the President's science advisor, reportedly attacked the deadlines as unwise. Though this pressure found some resonance among the

---

[1] The Nader study, *Vanishing Air* [1970], was written by a team headed by John Esposito. It provides a spicy account of events up through the 1960s.

House conferees, it was not sufficient to gain an adjustment of the deadlines. Muskie's time schedule was preserved, and 1975 and 1976 became the dates when the 90 percent reductions were to be achieved.

As a compromise with the House, however, Muskie and his Senate colleagues did allow a mild escape clause. The EPA administrator was granted authority to allow a single one-year extension of the deadlines, and the National Academy of Sciences (NAS) was mandated to conduct a study of the technical feasibility of meeting the standards and deadlines. Implicit was the thought that an NAS report which determined the emissions control program to be technically impossible under the tight deadlines would provide an occasion for an extension of the standards and probably for new legislation.

The result was the National Emissions Standards Act, enacted in December 1970 as Title II of the Clean Air Act. It established emissions standards for hydrocarbons (HC) and carbon monoxide (CO) for new cars in 1975, and an additional standard for oxides of nitrogen (NO_x) to be met in 1976. These new standards are to constitute a 90 percent reduction below the 1970 levels of CO and HC and the 1971 level of NO_x. (Administratively, these standards have since been specified as 0.41 grams per mile (gm/mi) HC, 3.4 gm/mi CO, and 0.4 gm/mi NO_x.) The Act also enables the federal Environmental Protection Agency to carry out certification tests to confirm compliance, to conduct research, to establish fuel regulations, to require and enforce warranties from auto manufacturers, and to certify and subsidize on-the-road inspection and testing programs. For violations of the emissions standards—marketing vehicles not certified—the Act imposes a fine of $10,000 per vehicle. This is the only enforcement device; in effect, it stops the production of model lines not certified.

The particular history of this legislation affects the current situation in two critical ways. There remains serious ambiguity about the intent of the Act, and there is great difficulty in stimulating the development and adoption of advanced automotive technologies. These problems can be traced to the forces that produced the 1970 amendments, and they are important elements of the context in which implementation of the Clean Air Act must proceed.

First, there is ambiguity about the intention of the Act in establishing a specific level of reduction in auto emissions. From one angle, the legislation can be understood as an effort by Congress and in particular by Senator Muskie and his subcommittee to exert pressure on the obvious political targets by means of legislation. The Congress was moved to do something forceful on behalf of the environment; and the automobile manufacturers, indisputably responsible and politically vulnerable, were an obvious target. The Act clearly "got tough" with them, by means of stringent standards, tight deadlines, provisions requiring them to supply data, and language requiring a good faith effort to comply regardless of the technical difficulties involved. The Executive Branch, controlled by the rival party, was another obvious target; it was vulnerable because of a sluggish record of enforcement. Congress embarrassed it by removing, through unusually explicit provisions, most of its administrative discretion in enforcing the Act. The overall tone of the legislation is that of an impatient Congress forcing a reluctant administrator and a resistant industry to act promptly.

Ironically, this is precisely the interpretation now favored by the automobile industry. Since the manufacturers are in no position to refuse to comply with the legislation, what they now want is to minimize the disruption to their normal business that compliance will entail. They have substantial and well-advertised efforts under way to develop control devices, and they clearly hope that these efforts, rather than actual achievement of standards, will be construed and accepted as compliance. In order to define compliance in this way, the standards must be interpreted as a first approximation by Congress, whose intent was not necessarily an actual 90 percent reduction but only achievement of the lowest emissions levels that are technically "reasonable." The preponderant technical opinion is that the technically reasonable rates would be less stringent than those associated with the objective of 90 percent reduction.

The opposing interpretation of the Act holds that the exact value of the 90 percent objective is of real social significance. The figure can be associated with calculations of the reductions necessary to ensure that ambient concentrations of air pollutants in all American cities will remain below levels associated with adverse health effects. These calculations are presented in the legislative history.[2] If the calculations are correct, then failure to meet the standards would damage the health of a significant element of the general population, and a relaxation of the standards would be much harder to justify.

The question of interpretation is destined to be hotly contested as implementation of the policy proceeds. Not only is the legislative history ambiguous, and the emissions reduction genuinely difficult technically, but the validity of the calculations used to justify the 90 percent objective is also open to question. The atmospheric processes that link emissions to levels of pollution in the air are as yet poorly understood. The effects of automotive pollutants at ambient concentrations are extremely subtle and inherently difficult to measure. Thus the data and the analytical models on which damage estimates must be based are not very robust, and conclusions cannot be established beyond valid scientific doubt. Moreover, the calculations were based upon the worst cases—Los Angeles and Chicago—which some argue yield figures that are too stringent for a national standard. The choice between these alternatives—affirmation of the 90 percent objective or an adjustment for technical convenience—will be one of the central issues of the developing program.

The second critical theme that emerges from the history of the emis-

---

[2] The specific figure of 90 percent was given in an analysis by D.S. Barth and others [1970] of the National Air Pollution Control Administration (now part of EPA). Their analysis showed that a 90 percent reduction was required to reduce ambient concentrations of CO, HC, and NO_x below those associated with adverse health effects. An analysis of the Barth work is provided in Chapter 7.

sions control program involves the development of automotive technology. In 1969, the Office of Science and Technology (OST) formed an Ad Hoc Panel on Unconventional Vehicle Propulsion to assess the possibilities of achieving low emissions with the ICE as compared with other technologies. The panel reported officially in March of 1970, concluding that modification of conventional engines was not a promising route to low emissions and recommending a federal program to develop an alternative technology to achieve emissions control [Office of Science and Technology 1970]. The recommendations of the OST panel were incorporated in President Nixon's message to Congress on environmental quality, delivered in February of 1970, in which he announced a new program of both governmental and private sector research on advanced low-polluting automotive technologies. The result was the Advanced Automotive Power Systems (AAPS) program, which became part of the newly organized Environmental Protection Agency. The program included funds for research directly commissioned by the government and funds for buying prototype vehicles developed by private parties.

In accord with the President's preferred time schedule, officials of the AAPS program laid out a plan to develop steam and gas turbine engines for production in 1980. In terms of this plan, the vehicles would be in early prototype stage by 1975, leaving five years to solve the numerous problems involved in reorienting production, marketing arrangements, service facilities, etc. Implementation of this plan, caught up in the larger politics of the issue, never really got started. The 1975 deadline for production vehicles, which Muskie's bill imposed, disrupted the time schedule and caused the program managers to change their orientation. No longer focusing on the development of advanced technology, the program was redirected to research on adjustments and minor modifications of the conventional ICE; it became another device to force compliance with the 1975 and 1976 standards. With this change the program lost coherence and purpose since, in the technical areas the EPA chose to explore, the manufacturers undertook much larger efforts on their own.

Furthermore, the AAPS program was not funded at a high enough level. Though authorized at a total of $55 million for a three-year period, actual appropriations have not exceeded $11 million in any year. Probably because it represents the President's initiative (and certainly because they do not trust the technical competence of the EPA), Congress has not shown any enthusiasm or generosity toward the program. Neither has the Office of Management and Budget, which has cut the program budget requests, arguing that it is the industry's business to conduct automotive research. The President, outflanked politically, has not pursued his initiative. The result is a nominal effort that can be cited as evidence of federal concern in response to pro-environmentalist political pressure, but which offers no serious possibility of developing a marketable alternative technology.

The conclusion of the OST panel remains, however: if the nation wishes to reduce automotive emissions by anything like 90 percent, then it had best change its automotive technology or at least contemplate such a change. This judgment reflects the workings of another major element of this problem—the state of the art in automotive engine technology.

THE CURRENT TECHNICAL SITUATION

At the core of the technical situation is a dilemma. The established technology— the ICE—which has evolved over the years into a successful power source for the mass market is inherently difficult to control to the 1975 and 1976 emissions standards. Technologies are known that offer much more promising means of achieving low emissions, but none of these can be incorporated in a light-duty vehicle at reasonable market prices on a three-year time horizon. Hence the combination of the strict standards and the tight deadlines discourages the adoption of alternative engines that are superior from an overall social viewpoint.[3]

Difficulties of Emissions Control

In essence, the operation of a standard spark-ignited ICE consists of a series of contained explosions. Because of the physical and chemical characteristics of combustion under these conditions, there are trade-offs among such engine performance parameters as (1) fuel economy, (2) emissions of CO and HC, (3) emissions of NO_x, (4) engine performance (i.e., acceleration and top speed), (5) vehicle driveability (including the propensity to hesitate, stall, or surge, and the relative ease of starting the engine), and (6) cost of manufacture. A desirable change in one dimension—say, a reduction of the NO_x emissions rate—will produce unfavorable changes in other parameters, other things being equal. Thus the strong, simultaneous controls on HC, CO, and NO_x emissions that the Clean Air Act mandates will involve inevitable and potentially severe costs in manufacturing, lost fuel economy, and reduced vehicle performance and driveability.[4]

These technical trade-offs strongly affect the emissions control package now planned by most manufacturers for their 1975 and 1976 models. In 1975, the controlled vehicles will include a system for leading some of the engine exhaust back into the carburetor (normally referred to as exhaust gas recirculation or EGR). They also will have an air pump and thermal reactor, as well as an oxidizing catalyst. In 1976, it is likely that a reducing catalyst will be added and the degree of EGR somewhat reduced. These devices must be tightly integrated with a number of engine modifications, such as changes in timing and air-

---

[3]An excellent review of the ICE emissions control problem and of alternative technologies is available in a report by the National Air Pollution Control Administration, *Control Techniques for Carbon Monoxide, Nitrogen Oxide, and Hydrocarbon Emissions from Mobile Sources* [U.S. Department of Health, Education, and Welfare 1970d].

[4]Indeed, a deterioration in vehicle quality is already evident from the controls imposed on the 1971-73 models, as can be seen in current advertising (particularly for gasoline) that plays to the motorist's displeasure with lost performance and driveability.

fuel mixtures, needed to produce the proper combustion characteristics. The operating parameters of the controlled engine must be held within narrower limits than are necessary for uncontrolled engines, and this requires more precise production tolerances. All these changes contribute to increased manufacturing cost.

With this particular control package, the performance of the vehicle is affected in a number of ways: EGR reduces peak combustion temperatures to control NO_x formation, but it tends to interfere with normal combustion and cause roughness, stalling, and stumbling of the engine. Additional fuel is added to the combustion mixture to counteract some of the undesirable effects of EGR, but this produces more HC and CO and increases fuel consumption. Thermal reactors and catalytic convertors, added to remove residual HC, CO, and NO_x from the exhaust stream, cause back pressures on the engine and reduce its effective power. As a result, the optimal settings of engine parameters for the purposes of emissions control are different from those that give the best fuel economy and engine performance. The controlled engine is more sensitive and more complicated than earlier internal combustion designs; its performance will be more likely to deteriorate, and it will need more maintenance. Because of the complexity of the systems, the maintenance will be more difficult.

The Stability Problem

The technical trade-offs, combined with the required tighter production tolerances and increased maintenance, make it likely that the controlled ICE will not be in stable equilibrium at the level of the 1975-76 standards, even if new test vehicles do pass the federal certification procedure. The performance of emissions control devices will tend to degenerate over time and distance traveled. The test procedures for certifying compliance with the 1975-76 standards do take into account the deterioration of control devices. (A study of federal test procedures is presented in Chapter 5.) Prototype vehicles must be run for 50,000 miles under specified conditions of maintenance and parts replacement and still perform within the standards. In effect, this means that automobile manufacturers must counteract intrinsic deterioration by achieving initial emissions rates somewhat lower than the standards. Unfortunately, however, the provisions of the federal certification test are not proof against the inherent dirtiness of conventional internal combustion engines. There are several reasons why vehicles will be *unstable* in the sense that on-the-road emissions rates will rise above those experienced in the prototype test:

1. The de facto rate of deterioration will be higher than that accounted for under the federal procedures because on-the-road driving conditions are more demanding and more damaging than those simulated in the prototype durability test. Currently, the 50,000 miles are accumulated by the manufacturers' own expert drivers during the summer months at 35 miles per hour on carefully maintained test tracks. In addition, the maintenance that vehicles receive in real life is not nearly as thorough nor as competent as that given the prototype test vehicles by the manufacturers' expert mechanics.

2. In the absence of elaborate government enforcement, few motorists will perform any specific air quality maintenance at all. Most vehicles will receive the maintenance necessary to make them run well, as they do now, and this will tend to increase emissions rather than reduce them.

3. Because emissions controls reduce engine efficiency and performance, there is a significant incentive for individual owners to disable, disconnect, or remove devices. Many motorists will do this kind of tampering, and the result will be greatly increased emissions.

The most significant element of the stability problem is the performance of oxidizing and reducing catalysts. If the 1975 HC and CO standards are to be achieved, the oxidizing catalysts must remove 80 percent or more of the pollutants in the exhaust stream. In order to meet current federal durability requirements, this efficiency rate must be maintained for 50,000 miles. Unfortunately, none of the oxidizing catalysts tested to date has proved durable over 50,000 miles of use. The catalysts with the highest removal rates do not even approximate the required durability. Thus the oxidizing catalysts will have to be replaced periodically, thus adding further burdens to the maintenance program.[5] For the reducing catalyst that may be used to control NO_x emissions, these problems are far worse.

In addition to these problems of normal operation, there are many conditions that cause the catalysts to fail completely. They can be deactivated by lead, phosphorous, or sulphur. Though lead is unlikely to be a problem at the manufacturer's test facility, it may present serious difficulties during the transition period on the road, when leaded gasoline will presumably have to be supplied for older vehicles. Moreover, even small amounts of lead, phosphorous, and sulphur now contained in "unleaded" gasolines can deactivate catalysts, and this poses a problem of potentially tight regulation of the composition of fuel and lubricants. Many oxidizing catalysts also can be destroyed by excess heat generated in their own operation by the reactions that oxidize HC and CO. Engine misfirings, rapid acceleration, heavy leads on the engine, or any other driving mode that puts a large amount of unburned fuel into the exhaust will heat the catalyst to temperatures at which many catalysts will be deactivated.[6] Once again, these events are unlikely to occur under prototype test conditons, but they are inevitable on the road. Some manufacturers are planning by-pass valves to protect the catalyst; others will simply allow the catalyst to deactivate.

Since ICE-powered vehicles are likely to be unstable, measures to enforce emissions limits on vehicles on the road must be considered. A massive inspection system might be set up to try to identify high-polluting vehicles and

---

[5]The EPA has allowed for one catalyst replacement during the officially defined 50,000-mile useful life of the vehicle. Thus catalyst material must last 25,000 miles under current rules. For NO_x catalysts particularly this is well beyond the current state of the art.

[6]The problems of catalyst durability are reviewed in the 1973 National Academy of Sciences study [1973].

send them back for mandatory emissions-control service or parts replacement. To do this on anything approaching a nationwide scale would be a staggering task, involving the individual inspection and regulation of tens of millions of individual vehicles. At present, few states have the capacity to run such programs[7]; and political opposition, inertia, inefficiency, and budget constraints will hinder their development. Inevitably, a substantial number of vehicles will escape effective regulation.

In short, projections of the effect of various emissions control policies must include some estimate of the instability factor. Forecasts based on the legislated maximum emissions rate are hopelessly optimistic. Likewise, it is not reasonable to assume that the stability problem will be easily solved by systems of in-use inspection and maintenance. The real difficulties of implementing such large-scale programs and the low likelihood of their even being tried in most states need to be taken into account.

Costs of Controlling the ICE

There have been several studies of the probable cost of attempts to meet the emissions standards for 1975 and 1976 with the conventional ICE. Table 2-1 summarizes the estimates of expected increases in manufacturing cost and fuel consumption. Looking first at new-car manufacturing cost, the estimated increase (over the cost of 1973 vehicles) to meet 1975 standards for CO and HC is between $160 and $214 per car. The net increase over the cost of an uncontrolled vehicle (shown in parentheses) is an additional $100, reflecting the controls imposed between 1967 and 1973. (The costs of controls imposed up through the 1973 model year are sunk costs and are not included in the analysis of future policy changes.) The cost to meet the 1976 standards (which include the stringent controls on NO_x) is between $270 and $400 per car more than the cost of 1973 vehicles. For an annual model run of 10 million cars, the total increase in manufacturing cost entailed by the 1976 standards could range up to $4 billion per year.

The increase in fuel consumption caused by 1975 and 1976 emissions controls has been variously estimated as indicated in the two right-hand columns of Table 2-1. Our analysis uses a range of values from 5 to 20 percent above the gas consumption of 1973 vehicles; it indicates that unless technical improvements are achieved, fuel costs will be increased by $1 to $4 billion per year by the mid-1980s.[8]

---

[7]To quote the National Academy panel: "In summary only a few states have any semblance of a testing/inspection system that would be adequate to ensure compliance in use [with the emissions standards]. Most states do not even have plans for such systems. The present service industry is inadequate to maintain the complex emission controlled hardware called for with the dual-catalyst planned for use in 1975-76" [1973, p.84]. This issue is discussed in greater detail in Chapter 6.

Table 2-1. Estimates of the Increase in First Cost and Fuel Consumption over 1973 Vehicles To Meet 1975 and 1976 Emissions Standards

Source	Additional Manufacturing Cost per Car		Increase in Fuel Consumption (percent)	
	1975 Controls	1976 Controls	1975 Controls	1976 Controls
National Academy of Sciences, 1972	$214 ($314)[a]		3-12	
National Academy of Sciences, 1973	$160 ($260)	$290 ($370)		(25)
Chase Econometric Associates, 1971	$164 ($247)	$269 ($352)		
Esso Research and Engineering, 1971				5-20
Chrysler Corp., 1971				(30)
General industry estimates		($245-400)		(10-26)

[a]Numbers in parentheses are estimates of increases over uncontrolled vehicles.

Only the roughest guess can be made about the actual economic costs of increased maintenance, but it seems reasonable to forecast a direct maintenance charge on the emissions control program of $10 to $20 per car per year—one-fourth to one-half of the cost of one additional visit to the garage each year.

If an attempt is made to handle the technical instability of the controlled ICE through a program of vehicle inspection and enforced service of air quality devices, this will add at least $3 per car tested. The mandatory servicing of control devices on the cars that fail the test will cost approximately $20 to $30 per vehicle. If the inspection is annual and roughly a third of the cars fail, then the enforcement program itself could cost over $1 billion per year.

When all these items are added together, the total cost of the abatement program is impressive. In the 1980s it will cost from $4 to $10 billion per year to control emissions from the conventional ICE.

Alternative Propulsion Technologies

The high cost and inherent instability of the tightly controlled ICE lead one to look for alternative means of propulsion that might be less polluting. A number of alternatives exist, the most prominent of which are the Wankel, the stratified charge ICE, steam and gas turbine engines, electric drive, and engines

---

[8]There has been a fuel penalty of 5 to 10 percent as a result of controls up through 1973, but again this is a sunk cost and is not included in the analysis. We assume a gasoline price (not including taxes) of twenty-five cents per gallon and a base fuel economy of fifteen miles per gallon.

that run on various types of clean-burning fuels. All of these, as well as various hybrids, are sufficiently different from the conventional ICE to require substantial redesign of vehicle components and production processes. Short of a major national commitment (much greater than anything suggested to date), these advanced designs cannot be prepared for mass production by 1976 or 1977. If intensified developmental efforts were to begin in 1973, however, any of these alternatives could be expected by model year 1981, with varying degrees of confidence in a satisfactory outcome.

The Wankel Engine. The Wankel engine is attractive to automobile producers quite apart from its emissions characteristics. For the power delivered it is smaller, lighter in weight, and has fewer moving parts than the standard ICE. Since the structural weight of a vehicle is related to the weight of the engine, a lighter engine translates into savings in manufacturing cost. Moreover, Wankel engines can be constructed in modules, so that engines of different size can have the same basic design, varying only in the number of rotors. The performance of the engine compares favorably to standard ICE's of contemporary design. The major problem is fuel consumption, which is currently quite high (30 percent greater than 1970 engines) and appears to be difficult to improve.

Unfortunately, the Wankel is not especially advantageous from the standpoint of emissions control. Because of its elongated combustion chamber, there is more quenching[9] of combustion than in a standard ICE, and hence greater concentrations of unburned HC in the exhaust. Formation of NO_x is slightly lower in the Wankel for the same reason and because a certain amount of exhaust gas is recycled in the operation of the engine. Additional EGR can be added but it tends to disrupt engine performance, and thus reducing NO_x to the 1976 standards is a problem. The Wankel will have to be carefully tuned to minimize emissions formation and, as with the standard piston engine, these adjustments may cause significant losses in fuel economy and road performance. Thus the Wankel engine will have many of the same stability problems that plague the ICE and will be even more dependent on the performance of catalysts and/or thermal reactors in the exhaust stream. If control devices fail, the resulting HC emissions from a Wankel will be greater than from a standard ICE.

Stratified Charge ICE. The Ford Motor Company, under contract to the Army Tank Command, is working on an engine that cleans up the internal combustion process by concentrating fuel around an internal ignition point and maintaining much leaner mixtures in the outer portions of the cylinder. The design of the cylinders and engine block is a significant departure from that

---

[9]Quenching refers to the phenomenon whereby combustion is retarded along the walls of the cylinder because they are cooler than the body of burning gas. Thus quenching stops combustion short of completion and leaves unburned hydrocarbons in the exhaust.

of the conventional ICE. Using catalysts, a prototype engine of this sort has achieved the 1975-76 emissions standards over short durations. The principle is also being applied to Wankel engines.

There are several problems with the Ford stratified charge engine that require further research. It requires very precise control of mixing in the combustion chamber, and this makes it more complex than the conventional ICE or the conventional Wankel. Both production and operating tolerances have to be tighter than for current vehicles. It would therefore be more costly to produce (potentially $250-$350 per car more than the controlled ICE). It also might be less stable in actual operation. Emissions would increase significantly as it wandered out of tune, and tuning for best emissions control may be different from tuning for best economy.

More encouraging results have been achieved by a Japanese manufacturer, Honda. The Honda design is different from that of the Ford engine because it involves a small chamber with an extra valve opening onto the main combustion chamber for each cylinder. A rich mixture is fed into the small chamber by a separate carburetor, and ignition takes place there. Combustion then spreads to the cylinder chamber which contains a lean mixture. This process avoids the need to inject fuel under pressure and to control fuel mixing, and thus it operates under greater tolerances and would be less costly. For example, in small cars Honda engines meet the 1975-76 standards at low mileage with no catalysts or EGR. The Honda engine has drawn high praise from the National Academy of Sciences panel for its inherent technical characteristics, and it offers the advantage that it would require only a relatively modest adjustment in production processes. There are problems as well. The engine is relatively heavy for the power it delivers and to date has been demonstrated only in prototypes smaller than most cars on the American market. It is not clear what problems will be ecnountered in scaling it up to larger vehicles; higher horsepower versions may entail fuel penalties and the need for control devices in the exhaust stream to hold to the standards.

Rankine Cycle. Rankine cycle engines use continuous combustion to heat a working fluid (water or some organic solution) which then expands against pistons (or a turbine) to provide the work required to propel a vehicle. This technology was used in early motor vehicles—e.g., the Stanley Steamer— but it has not been developed for the modern mass market. Such an engine has several attractive features. Compared to the ICE, it produces high torque at low engine speeds and thus does not require a complicated transmission. It can run on low-octane gasoline as well as less refined fuels, such as kerosene. Because the fuel is not exploded, the Rankine is a quieter engine. It is also extremely durable and would require less maintenance and, presumably, less frequent replacement than an ICE. There are no barriers to mass production, and it is likely that such engines could be prepared for the automotive market at cost roughly equal to the tightly controlled ICE ($1,200 to $1,300 per unit, or about 30 percent more than the cost of current engines). Fuel economy (ten to fifteen miles per gallon) promises to be comparable to the current ICE.

Rankine engines offer stable low emissions. Combustion is more complete and thus the engine exhausts far less HC and CO than an ICE. Also, the gases cool less rapidly after combustion, so less NO_x is formed. Even without special efforts to control emissions, experimental Rankine engines have produced emissions of less than 0.2 gm/mi for all three pollutants—less than half of the 1975-76 standards for HC and NO_x and a tenth of the 1975-76 standard for CO.

The traditional disadvantages of the Rankine engine—freezing of the working fluid in cold weather, slow start-up, danger of boiler explosion, use of scarce materials, and sheer bulk—have all been either eliminated or greatly reduced even with the small amount of research done in recent years. Organic working fluids now being tested freeze only at $-30°F$. There is no boiler to explode, and if a leak should develop in the tube where the working fluid is heated, not enough would escape to constitute a hazard. Start-up time is now down to under one minute, and it is doubtful that this would prove much of an inconvenience, particularly since starting would be more reliable than for a controlled ICE. At the moment, Rankine engines are heavier than the ICE, but this disadvantage is reduced because the transmission is lighter than the transmission in an ICE-powered vehicle. Recent advances in condensers, heat exchangers, and expanders have reduced the volume of the engine, which will now fit into the smaller model lines currently on the market (e.g., a Ford Fairlane or Chevy II).

The major disadvantage of the Rankine, as with any radical alternative, is that such engines are not now in use, so the technical momentum of the industry works against them. Even if Rankine technology were universally acknowledged as superior (which it is not), there would be difficulties in overcoming the well-established commitment of the industry to the ICE. Though their durability and emissions characteristics may make Rankine cycle engines attractive from the point of view of the motoring public, the manufacturers, faced with costs of reorganization and retooling, are inevitably less enthusiastic.

Gas Turbine. The gas turbine engine is an automotive adaptation of the technology commonly used in jet aircraft. It has very attractive thermodynamic properties and yields power and energy densities (the determinants of speed and range) greater than those of the ICE. Because of its thermodynamic characteristics and because the aircraft industry has given it an independent (and conceivably competitive) base, the gas turbine has long received a significant amount of attention from the automobile manufacturers. Apart from the ICE, it is the alternative with which the industry is most familiar and the one (at least until the advent of the Wankel) to which it has been most favorably inclined.

In terms of emissions control, the gas turbine is inherently cleaner than the ICE and much more stable. A properly tuned turbine engine produces HC and CO emissions in the range of the 1975 standards, and these rates do not increase significantly over the life of the engine. There is a problem with NO_x, however. Although NO_x formation is lower than in the uncontrolled ICE (around 2 gm/mi), it still is significantly higher than the 1976 NO_x standard, and there is no control technique at the moment which will reduce it. Of course, even with higher initial emissions rates, the gas turbine might well produce less pollution than the controlled ICE because of its greater stability.

There are barriers to the mass production and mass operation of gas turbines in automobiles. They require exotic materials and more tightly controlled production processes than the automobile companies have developed to date. The manufacture of turbine engines also would require drastic changes in the current labor force, involving either retraining or replacing many workers. There would be similar problems in the maintenance industry: the turbine could not be handled by current garage mechanics. The unit price of the engine is hard to predict, but it probably would be significantly higher than the controlled ICE ($1,500 or so by current estimates).

Electric Drive. Recent research has identified new battery technologies that have sufficient power output and energy storage to permit an automotive application. If fully developed, such batteries might power a passenger car with the speed, range, and performance of current ICE's at a competitive cost. Such a vehicle would have no HC, CO, or NO_x emissions whatsoever and would be quieter, smoother, and more durable in operation. The pollution problem under such a technology would be displaced to a smaller number of larger sources (i.e., electric power generating stations) where it is reasonable to expect significant economies of scale in pollution reduction.

Such vehicles also would make possible a more flexible energy policy and would help to conserve petroleum. At some point in the future, fossil fuels will become too expensive to use for propelling individuals around in 4,000-pound containers. Naturally, there is great debate about how far in the future this point lies and whether it is desirable to effect such a change at an earlier date in order to reap the benefits of reduced pollution.

It would require a substantial (but by no means unattainable) development effort to prepare electric drive technology for mass production by the early 1980s. Though such an effort is unlikely to emerge under current market conditions, it would be possible with government financing. The more serious barriers to electric drive technology are its broader economic consequences. The reallocation of the productive capacity of the automotive manufacturers would be much more substantial than in the case of the Rankine cycle, Wankel, or stratified charge engines. Disruptive effects would be concentrated in markets for certain materials and parts. Oil producers and distributors would lose their gasoline market, which currently is the major portion of their business, although this would occur at a gradual pace determined by the turnover in the vehicle population.

Electric utilities would experience a greater increase in demand than they are now estimating, and they would have to undertake greater capital investment. This again would be a gradual process phased to the vehicle turnover rate.

The essential question, then, is whether long-run economic forces favoring electric technology will appear soon enough, or whether the gains from centralizing emissions sources will be significant enough, to make a battery-powered car competitive within the time frame of current policy (fifteen to twenty years). This is a matter of great uncertainty. But since the decisions on automotive technology now being forced by the Clean Air Act will strongly affect the patterns of energy use until at least 1990, the issue ought to receive serious and immediate attention.

Internal Combustion Engines with Gaseous Fuels. It has been demonstrated that dramatic reductions in exhaust emissions from the conventional ICE can be achieved by changing to gaseous fuels such as liquefied petroleum gas or compressed natural gas. Such systems, which require only moderate changes in engine design, have inherently clean combustion and good performance characteristics, and they result in stable emissions near the levels of the 1975-76 standards. Some taxi fleets and pools of government vehicles have been converted to gaseous fuels in recent years, often with net savings in vehicle cost.

The main problems with gaseous fuels concern the supply system and safety.[10] A major increase in the national supply of gaseous fuels would be required (perhaps through coal gasification or imports of liquefied petroleum gas) in order to serve all the vehicles in the country. Even if the supply were available, the widespread use of gaseous fuels would require a major overhaul of the distribution system for automotive fuels. These two factors tend to limit the potential use of gaseous fuels to fleet operations in large cities. The safety problem appears to be more a matter of public acceptance than of real danger. Most analyses of the safety question conclude that a properly run gaseous fuels system is as safe as one based on gasoline, if not more so.

A more adventuresome approach now under study would involve the use of hydrogen as an automotive fuel. The hydrogen could be made from hydrocarbon fuels or from water (using electric power as an input) and delivered in liquid form much as gasoline is today. Or it might be manufactured from a conventional hydrocarbon fuel within the vehicle itself. A hydrogen engine would be essentially pollution free.

As with other alternatives mentioned above, additional research is needed to determine if these new approaches are practical for mass-produced cars. One thing is clear, however. Technologies are available that can be expected to

---

[10]A very useful analysis of gaseous fueled vehicles has been published by the Environmental Quality Laboratory at the California Institute of Technology (see *Smog—A Report to the People* [Lees 1972]).

solve the stability problem and that could be developed for new car production by, say, the model year 1981. It is unrealistic, however, to think that any of these potential solutions will emerge in a timely fashion from the current emissions control program.

THE ISSUE OF EXTENDING THE STANDARDS

Since the major element of discretion left to the administrator by the Clean Air Act is his ability to grant a one-year extension of the 1975 and 1976 standards, much of the debate over the program during the first three years of its operation has focused on that question. In terms of its direct impact on the quality of the atmosphere the question is insignificant. The reductions achieved by the program accrue gradually over a ten-to-fifteen-year period, and the effect of a single year's delay is far outweighed by uncertainty over what emissions rates will actually turn out to be. The issue has served, however, as a symbol of government resolve and a focus for industry complaint.

In February of 1972 all the major manufacturers filed for an extension of the 1975 standards, arguing at length that the catalyst-based systems that they were developing could not be expected to perform on mass-produced vehicles at the emissions rate required. They argued that they needed at least the extra year allowable for further development. After mandatory hearings on the subject, the then EPA administrator Ruckelshaus denied the requested extension on the grounds that the companies had not compellingly demonstrated that the stringent conditions set forth in the act for granting an extension had in fact been met. The act requires a one year's extension of the standards be granted only if it is shown that: (1) it is in the public interest, (2) all good faith efforts to comply have been made, (3) technology is not available to the applicant or anyone else, and (4) the National Academy of Sciences panel concurs that appropriate technology is not available. Thus an applicant for an extension of the standards carries a heavy burden of proof, and Ruckelshaus ruled that the manufacturers had not met it.

It is reasonable to surmise that for both the industry and the EPA administrator the immediate issue was not the primary concern. Ruckelshaus, in his first significant test on the mobile source area, needed a show of strength, a signal to the industry that EPA would not lightly yield. The manufacturers, heavily committed to what was proving to be an unworkable technology, needed to write a public record of their difficulties. Down the road loomed the possibility of a major confrontation; if the industry could not certify the only vehicles they could manufacture then the only recourse of the government was to impose the $10,000 per vehicle fine, thus halting automotive production. This would be a modest version of that central nightmare of the nuclear age—a situation in which a border incursion must be resisted either with the hydrogen bomb or not at all.

In the interests of elaborating the record and still in hopes of postponing commitment to the use of catalysts, the industry sought to have the administrator's decision reversed by the United States Court of Appeals. In December of 1972, the District of Columbia Circuit Court ordered Ruckelshaus to review his decision, explicitly taking into account the *Semiannual Report* of the NAS Committee on Motor Vehicle Emissions which had been issued the previous January [1972]. When the administrator reaffirmed his denial of a delay, the court ordered a new set of hearings and a new decision, directly stating that the weight of evidence justified a delay of the 1975 standards and assigning to the administrator the burden of proof in denying such a delay. The hearings mandated by the Court produced more testimony on the unpromising prospects of the catalyst-based technology and on the difficulties in preparing it for the 1975 model year. A second report by the NAS panel issued in February of 1973 vaguely pronounced the catalyst-based systems a technically feasible approach to the 1975 standards but bluntly labeled them the "most disadvantageous" of the available options [p.5]. Against this backdrop Ruckelshaus relented. He granted a year's delay in the national standards for 1975, set interim national standards which represent about a 50 percent reduction below 1970 vehicle emissions (as opposed to the 90 percent called for in the act), and set a more stringent interim standard for California (about a 75 percent reduction below 1970 vehicle emissions).

This decision, which the industry decided not to appeal, complied with the suggestions of the District of Columbia Circuit Court but did not provide much relief to the manufacturers. Under the interim California and national standards, the industry still must be prepared to use catalysts on approximately 20 percent of their production run and thus did not avoid the necessity of investing in that technology with the required expenditure of catalyst fabrication plants and the like.

At the date of this writing the elaborate process grinds on against the warning of at least part of the NAS panel. Some members of the panel, the NAS report said,

> are concerned that strict enforcement of the provisions of the Act might, by enforcing adoption of the control system first to be developed and certified, defeat the goal of the earliest possible attainment of compliance by the most generally desirable means ... [some] believe that, once having embarked on large-scale production of the catalyst-dependent control systems, several years would elapse before major manufacturers would alter course in favor of producing more generally satisfactory vehicles. [1973, p. 125]

Clearly it is time to consider options.

---

[11] In July 1973, a one-year extension of the 1976 NO_x standard was also granted, and an interim standard was set representing a 50 percent reduction of 1971 emission rates.

Chapter Three

Policy Options and Predicted Outcomes

Henry D. Jacoby
and
John D. Steinbruner

A fertile imagination can generate a large number of different approaches to the automobile emissions control problem. These include an outright ban on the ICE, myriad systems of effluent taxes, enforced maintenance, gasoline rationing, traffic controls, regional rather than national programs, and a substantial reduction in the abatement objectives. A full evaluation of the options would require an extensive discussion indeed. As a practical matter, though, there seem to be three basic options, illustrated in the top half of Figure 3-1. One option is to continue with the established policy of full implementation of the Clean Air Act as it now stands. In this case there is a subsidiary choice about how hard to push for enforcement of emissions performance "on the road." We shall refer to this option by the shorthand notation of EST, or, if an enforcement program is included, as EST/ENF.

A second option is to relax the current emissions standards to technically convenient levels in order to mitigate the cost and stability problems just discussed. This we shall refer to as option REL. And the third option, which we call ALT, is to adopt a vigorous program to develop one or more of the low-polluting propulsion technologies and to prepare for its adoption in the early 1980s. In this last case there is a subsidiary choice about what to do about emissions standards while this alternative is being prepared. On one hand, it is possible to hold to established standards through the late 1970s (option EST/ALT); or else standards can be relaxed in the interim, to be raised again in the early 1980s (option REL/ALT).

No matter which path federal policy takes there is a wide range of possible outcomes, as displayed in the bottom half of Figure 3-1. First, there is the question of program cost over some period (we use fifteen years); our analysis will utilize high, medium, and low assumptions about the costs of emissions control. Then, whatever the cost turns out to be, after some years of experience with vehicle deterioration under road conditions there will be some

28 Clearing the Air

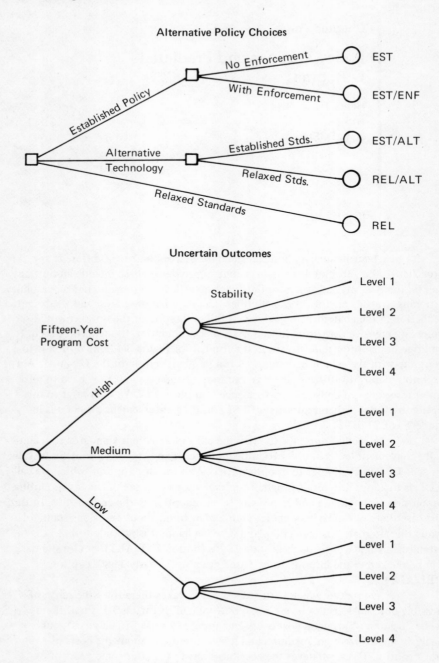

Figure 3-1. Schematic View of Alternative Policy Choices and of Uncertain Outcomes Under Each of the Policies

realized level of emissions stability. As the figure shows, the analysis will incorporate a range of assumptions about the degree to which vehicle emissions deviate from those predicted by the prototype test. The relative attractiveness of different policy options, naturally, depends on where they are expected to come out on the diagram of cost and stability.

ESTABLISHED POLICY

When understood in the context of the 1960s, the provisions of the Clean Air Act have a simple, appealing logic. The Act gets tough with the oligopoly of manufacturers who are widely believed to have colluded to fend off California's attempts at emissions control. The rigidly imposed standards and deadlines, the clause requiring good-faith efforts to meet them, and the threat of prohibition from the market are all devices for breaking the resistance of recalcitrant, profit-oriented industrialists. Explicit constraints on auto emissions are given the force of law. Enforcement of explicit standards requires measurements which must be recorded and made part of the public record, providing a political focus and a pressure point for diffuse and weakly organized environmentalist forces. Moreover, the difficult task of designing control systems is left to the manufacturers, who, so the argument goes, have an incentive to produce the most efficient technical solution. State inspection programs, along with the warranty provisions of the Act, should prevent manufacturers from producing a vehicle that turns into a bad polluter after a few months in the owner's hands. The implementation of this logic can accurately be referred to as the "established policy" of the Congress and the Environmental Protection Agency (EPA).[1]

There is some chance that this approach may produce an efficient method of controlling ICE emissions. Perhaps there will be breakthroughs in technology and production techniques yielding dramatic reductions in emissions at low cost with no undesirable side effects. (Catalyst manufacturers are in effect predicting such a breakthrough when they forecast the development of cheap, durable, highly efficient catalytic devices.) Such an outcome would be an extraordinary bit of good luck and would justify the established policy. The technical momentum of the industry and the political forces behind the current policy would be directed down the right track, and at least one environmental problem would be solved.

Unfortunately, if the technical breakthroughs are not so easily achieved, then the established policy is much more problematic. Despite official neutrality about propulsion technology, the 1975-76 standards and deadlines serve to lock the entire industry into a narrow range of options. Routine production procedures in the industry require that basic engineering designs be established three years be-

---

[1] There are important matters that to date have not been decided, including elements of certification testing and assembly-line surveillance, and guidelines for vehicle maintenance and state enforcement.

fore production begins (July 1972 for the 1975 models). This timing allowed only eighteen months between the enactment of the 1970 amendments (the stringency of which was not anticipated by the industry) and the onset of the production cycle for 1975 vehicles. The tight deadline could not help but solidify the commitment throughout the industry to the conventional ICE and to "bolt-on" devices.

If the technical trade-offs hold in this situation, then the response of automotive designers, working under time pressure, is predictable. They will aim for a vehicle design that (1) will pass the federal prototype test so it can be sold and (2) will entail the least possible increase in new car cost and the smallest losses in vehicle driveability. To achieve this will require compromises with vehicle emissions performance over long periods of time and miles of use. Vehicle emissions will tend to be "unstable" in the sense that, without continuous and informed maintenance, emissions rates will rise significantly under the conditions encountered on the road.

Emissions Forecasts

Just how high emissions might go is a matter of considerable uncertainty. The factors contributing to instability have been identified above; the degree to which they operate will depend both on vehicle design and the pressure of government enforcement. In order to analyze these phenomena we have developed a set of models of the automobile population, taking into account growth, turnover, and use patterns as well as the differing emissions standards in force in different years. They also contain submodels of vehicle deterioration under road conditions and of auto emission inspection and enforced maintenance.[2]

The key to the analysis, as noted earlier, is the concept of "stability" of vehicle emissions over time. A vehicle type is stable if it will, on the average, perform near to the prototype test result even without elaborate enforcement systems and specialized emissions control maintenance. The greater the increase in emissions without rigorous enforcement measures, the more unstable a particular design is considered to be. We define four stability levels that cover the range of emissions performance that could result from current efforts to control the ICE. There is also a stability class designed to represent a stable alternative technology. Figure 3-2 shows a simplified version of this part of the emissions model. Each level of stability indicates a different state of the world that might result from the inherent properties of the vehicle design, voluntary maintenance, the driving habits of the population, and the other factors affecting emissions control system deterioration.

Level 1 stability reflects what might be achieved if vehicle design and production were at the most favorable limits of imaginable success and if vehicles were voluntarily maintained exactly according to manufacturers' specifications. It presumes a catalyst that would not fail completely even under high temperature, rough driving, or poisoning. This level of stability is very unlikely to occur, but it is conceivable and thus defines the lower end of the range. Level 4

---

[2]See Chapter 6 for a more detailed discussion of the models that lie behind this analysis.

stability is the high end of the range. It presumes a catalyst that is prone to failure and a significant amount of perverse maintenance (adjusting or removing of control devices in order to improve performance at the cost of higher emissions). Further, it assumes that whatever air quality maintenance is done will be inexpertly performed. Levels 2 and 3 represent intermediate cases to provide reference points within the established range. Level 0 stability indicates the performance that could be expected from a stable alternative technology.

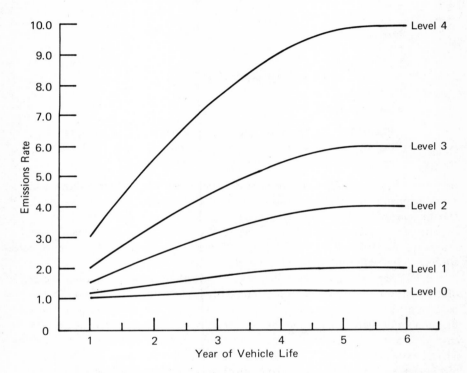

Figure 3-2. Emissions Rates for Different Stability Levels on the Assumption That No Specific Emissions Control Maintenance Is Performed, Stated as a Multiple of the Prototype Test Result

Using the emissions model and these definitions of stability, we can prepare alternative emissions forecasts. Taking CO as an example, Figure 3-3 shows the annual total emissions for the nation for each of the fifteen years between 1975 and 1989. The lowest curve in the figure duplicates an estimate prepared in 1972 by the National Academy of Sciences[3] and assumes perfect stability in emis-

---

[3] See the *Semiannual Report* by the Committee on Motor Vehicle Emissions to the EPA [1972, pp. 46-47 and 76-77].

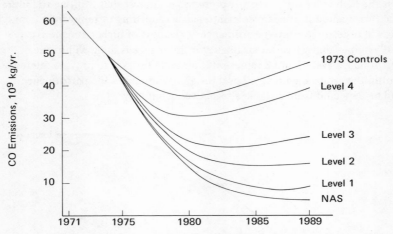

Figure 3-3. Nationwide Carbon Monoxide Emissions for Alternative Vehicle Stability Levels Under Established Policy, With No Program of Inspection and Enforced Maintenance

sions (that is, zero deterioration of post-1975 vehicles). The top curve shows what would happen if emissions controls were left at 1973 levels. Other curves show total emissions with established policy (EST with no enforced maintenance) under alternative assumptions about stability. For example, the curve labeled Level 4 results from established policy if, in fact, vehicles produced after 1975 deteriorate according to the definition of stability Level 4 in Figure 3-2.

Several things about Figure 3-3 are worth noticing. First, the effects of different policy options do not differ greatly in the early years (say, 1975-79) because of the gradual turnover of the vehicle fleet. It takes several years before the number of tightly controlled vehicles grows to be a significant portion of the auto population, and the results of any emissions control program play out over a decade or more. Another point to notice is that unless emissions controls are highly stable, total emissions will be on the rise again in the late 1980s due to the increase in total miles driven. Finally, the figure shows that only the most optimistic assumptions (Level 1 or better) approach the goal sought in the Clean Air Act of a 90 percent reduction in emissions.

To the results for CO shown in Figure 3-3 must be added similar data for HC and NO_x, and for analysis of broad policy options these complex phenomena must be aggregated into an understandable set of summary indices. We aggregate pollutants by taking a weighted sum of total national emissions of CO, HC, and NO_x in each year.[4] Aggregation over time is accomplished by taking

---

[4]The weighting factors that lie behind the calculations presented here are CO = 0.12, HC = 1.0, and NO_x = 1.0. The analysis was conducted using three distinct sets of weights, and the central conclusions proved insensitive to the weighting scheme used. See Chapter 6 for details.

the simple sum of the weighted emissions over the fifteen-year period of 1975-89. The result is a set of estimates of the total tons of "pollutants" spewed out by cars under different policy options. Since we are concerned with the *incremental* reduction associated with the tightening of standards programmed for 1975 and 1976, we take the pollution that would be experienced under 1973 controls as the base condition. The percentage cutback achieved by a particular policy (below levels produced by 1973 controls) is termed its "Weighted Index of Reduction." This is our measure for comparing the effectiveness of the different policies laid out in Figure 3-1.

The values of the Index for established policy are in the first column of Table 3-1. (The values for other policies are there as well; we shall return to them shortly.) As the table shows, emissions may be reduced, below those from 1973 controls, by as much as 55 percent or as little as 13 percent, depending on the stability level attained.

Table 3-1. **Weighted Index of Reduction in Emissions Under Different Policy Options, for Indicated Stability Levels for ICE Vehicles and Level 0 for Alternate Technology**

Stability Level	Weighted Index of Reduction, 1975-1989				
	EST	EST/ENF	REL	REL/ALT	EST/ALT
1	.55	.59	.37	.51	.60
2	.45	.54	.11	.37	.55
3	.35	.47	−.15	.23	.49
4	.13	.34	a	a	.38

[a]Level 4 is not applicable under relaxed standards.

The significance of such percentage reductions is not intuitively obvious, of course, for they are not measures of the things we value about pollution control—such as the health of human beings, plants, and animals.[5] One can, however, make very rough estimates as to what these reductions imply for current efforts to meet established ambient air quality standards. In a city such as Philadelphia, for example, where automobiles in 1967 contributed about 60 percent of CO pollution, an emissions control program experiencing poor stability (Levels 3 or 4) might result in ten to twenty periods each year (i.e., one to two days per month) during which the eight-hour standard for CO concentrations was exceeded.[6] If the ambient standards are accurately set, then adverse health effects would still be suffered even after fifteen years of the control program.

---

[5]A quantitative approach to this problem is provided in Chapters 7 and 8.

[6]This would occur in 1985-89 assuming Level 3 or Level 4 stability under the established policy, normal weather and traffic conditions, and a 50 percent abatement of stationary source pollution. For further details of such calculations see Chapters 7 and 8.

The Role of Enforced Maintenance

Results like these just described, which show the effect of technical instability on the degree of pollution eliminated, inevitably suggest the potential value of a program of inspection and enforced maintenance as a means of controlling the deterioration of emissions controls. But the difficulties of establishing such a system on a national scale are formidable. Fifty states and countless local jurisdictions would be involved, few of which now have the legal authority to establish inspection and maintenance programs, much less the capacity to make them work.[7] Moreover, to build a program capable of controlling an unstable technology not only would require setting up (or supplementing) testing and inspection programs in twenty or thirty states at a minimum, but would also require regulation of maintenance facilities, licensing of mechanics, etc. And since all of this would entail significant harassment of motorists, such a program would be politically hazardous. In short, it is questionable that a comprehensive enforcement system can be created, or if it were created that it would operate efficiently.[8]

Still, a maintenance program is an important part of control policy as conceived in the Clean Air Act, and it deserves careful attention. There are many ways such a program could be organized—some more effective than others in reducing emissions. The format most likely to be adopted involves state-run inspection stations which try to identify high emitters and send them to certified private garages for restoration of their air quality control systems.

California and New Jersey are the only states to date that have attempted inspection programs. Following the experience with these programs, most states will probably set the allowable emissions rate so that some predetermined percentage of all vehicles will fail the inspection. No state administration wants a system that fails so many cars as to be politically unacceptable or so few that it appears to be a waste of time. Furthermore, the test criteria are likely to be designed so that the percentage failed is roughly the same for all model years. It would be hard to sustain a program that penalized the owners of old cars and let new-car owners go free—or vice versa.

Based on these assumptions about what states will in fact do if they

---

[7] The NAS panel made a review of state inspection and maintenance capacity and reached a pessimistic conclusion [1973, p. 82-84]. Many of the specific difficulties are discussed in Chapter 6.

[8] It is argued that enforcement ought to be selective by local area according to the severity of the air pollution problem. Montpelier, Vermont, would allow its motorists to gain the benefits of lax controls, while Los Angeles would impose a vigorous enforcement program, perhaps even requiring a different vehicle design. Regionalization is attractive if there are only a few areas requiring stringent controls, but unfortunately Los Angeles and Philadelphia are not the only cities with serious problems. A study by the Office of Science and Technology concludes that 70 percent of the vehicle population would require controls. See *Cumulative Regulatory Effects on the Cost of Automotive Transportation (RECAT)* [1972]. If the vehicles that must be rigidly controlled constitute a significant portion of the population (say, 40 percent or more), then this analysis of a national program holds for a regionalized program as well.

are induced to set up inspection and service programs, an evaluation has been prepared for a national system that fails 30 percent of the cars each year. Cars receiving the mandated service under this program are considered to be restored to the level of emissions control they exhibited as new vehicles. That is, the emissions of cars failed are forced back to the lowest point on the appropriate stability curve in Figure 3-2.[9] The assumption of such effective maintenance is extremely optimistic; it obviously represents the outer bound of what can be expected.

This version of the established policy with enforcement on the road is given the shorthand name EST/ENF, and the Weighted Indices of Reduction under this option are shown in the second column of Table 3-1. These data, which present a picture purposefully biased in favor of maintenance programs, illustrate the fundamental dilemma inherent in this policy instrument. Under highly stable conditions, when maintenance cannot produce a substantial reduction in automotive emissions because there is none to be had, an enforcement program does not increase the effectiveness of the control program. Under low stability conditions (say, Level 4), enforced maintenance does increase effectiveness, but the ultimate improvement over 1973 controls is still not very great.

Program Costs

Estimating the costs of the different policy options for emissions control is, of course, an exercise in making reasonable assumptions. Until controlled automobiles are mass-produced and driven under road conditions, the actual cost of emissions control cannot be known. Even then, accurate cost calculations will require an elaborate effort not likely to be undertaken. It is important, therefore, to adopt an analytical framework that can accommodate this fact and to work with a range of possible estimates reflecting the inherent uncertainty.

The discussion above identifies four components of cost: (1) the increase in vehicle manufacturing cost; (2) the increased cost of vehicle maintenance; (3) the decline in fuel economy; and (4) the cost of inspection, specialized air quality control service, and general administration necessary to enforce performance on the road. Table 3-2 presents low, medium, and high estimates of the fifteen-year total of these costs under current policy. The low estimate assumes an increment in manufacturing cost over that of 1973 vehicles of only $100. This is extremely optimistic—well below the current official estimates shown in Table 2-1. The estimated fuel penalty, at 5 percent, is also at the low end of most estimates.

The medium cost estimate reflects the results of most of the official

---

[9]The level of stability is assumed to be inherent in the technical design and thus cannot be changed, though a maintenance program can hold actual emissions rates below the maximum values. It is possible, of course, that harassment of motorists on the road might reverberate back to the design labs in Detroit, and that a rigid enforcement program might lead to more stable vehicles at some point in the future. It is argued, however, that this linkage is very weak.

government studies of control system cost. The high estimate is consistent with most industry analyses of manufacturing expenses and with the more pessimistic predictions about fuel economy.

Table 3-2. Low, Medium, and High Estimates of Fifteen-Year Program Costs Under Established Policy at a 5 Percent Discount Rate

Cost Assumptions	Fifteen-Year Program Cost ($ Billions)	
	No Enforced Maintenance (EST)	With Inspection and Enforced Maintenance of 30% of Vehicles (EST/ENF)

Low cost
  Initial cost = $100/car
  Regular maintenance = $10/car per year
  Fuel penalty = 5%
	27.4	40.5

Medium cost
  Initial cost = $250/car
  Regular maintenance = $20/car per year
  Fuel penalty = 10%
	61.5	74.6

High cost
  Initial cost = $400/car
  Regular maintenance = $20/car per year
  Fuel penalty = 20%
	94.0	107.1

One implication of these estimates is immediately evident. If $30 to $100 billion is to be spent on controlling automotive emissions, then the government surely ought to think broadly about its policy options. With that amount of money at stake, even very substantial changes in the current manufacturing process can be considered; even very risky investments in technical development can be justified.

Possible Failure of Prototype Tests

In addition to the problems of stability and enforcement, there is a significant possibility that the automobile manufacturers simply will not be able to pass the prototype test with their 1977 model vehicles. To date, none of the U.S. manufacturers has perfected a system which meets all three standards under the established certification procedure. Should this situation continue into 1975, given that one-year extensions have been granted, then there will be a clear confrontation over the established policy. The government will be legally committed to the standards; the industry will only be able to market vehicles which do not meet the standards. If it comes to that impasse, some adjustment of the policy will be required, for no government could stand the pressure resulting from a halt in production by even a single manufacturer, much less by the whole industry. The negative economic effects of such an event would completely outweigh the benefits to be gained from pollution reduction, and the political response would be commensurate with the stakes at hand.

RELAXED STANDARDS

Some observers look at the manufacturers' complaints about cost and technical difficulty and the risks for the government, and conclude that the standards should simply be relaxed. There is uncertainty and hence disagreement about the precise degree of relaxation needed to avoid these problems, but a reasonable approximation would set new standards at 8.5 grams per mile (gm/mi) CO, 1.0 gm/mi HC, and 1.0 gm/mi NO_x. These standards would represent a 75 percent reduction in emissions rates, as compared with the 90 percent called for in the Act. This option is denoted as REL in Figure 3-1, and the associated values of the Weighted Index of Reduction are presented in Table 3-1.

Such relaxed emissions rates probably could be met with control techniques currently under development, and the trade-offs would not be as severe as under current standards. Though higher at the outset, such emissions rates would be more stable. There would be less EGR, and no reducing catalyst. There might be an oxidizing catalyst, but with a lower required efficiency rate it should be more durable. Other adjustments would be less extreme, allowing designers to retain vehicle performance, driveability, and fuel economy—and at reduced cost.

The degree of cost reduction is uncertain, so once again we calculate three cost estimates that span the range of likely outcomes. Table 3-3 shows a

Table 3-3. Low, Medium, and High Estimates of Fifteen-Year Program Costs Under Relaxed Standards at a 5 Percent Discount Rate

Cost Assumptions	Fifteen-Year Program Cost ($ Billions)

Low cost
  Initial cost = $75/car
  Regular maintenance = $5/car per year
  Fuel penalty = none
	13.9

Medium cost
  Initial cost = $150/car
  Regular maintenance = $5/car per year
  Fuel penalty = 5%
	30.2

High cost
  Initial cost = $250/car
  Regular maintenance = $10/car per year
  Fuel penalty = 10%
	53.7

low estimate of incremental cost—only $75 per car over 1973 models and no fuel penalty whatsoever. The high estimate resembles the medium estimate for the existing policy. This is not unreasonably high, since most of the equipment required to meet established standards is needed to satisfy the relaxed emissions ceilings.

A central argument for relaxation of the standards asserts that most of the benefits of automotive emissions control can be achieved with a lesser reduction than that sought in the current legislation. Thus, it is argued, even if the relaxation of standards does yield higher emissions, the resulting increase in air pollution levels will not be very damaging—especially when balanced against substantial cost savings.[10]

Though a relaxation of the standards would draw strong political criticism, a serious case can be constructed in its defense. After utilizing the tight standards of the Clean Air Act to drive the industry to commit itself to emissions control, this policy then retreats, setting new standards based on the information made available by the latest research. Such an adjustment can be portrayed as a reasoned balance of competing claims. And, by enabling all manufacturers to pass the prototype test, this option avoids the danger of provoking a sharp political confrontation on the issue.

The problems with relaxed standards arise from the fact that such a policy would stabilize pollution at higher levels than the current approach. Unless gains in stability from the relaxed standards were dramatic, they would not compensate for the higher emissions rate. Moreover, in relaxing the pressures on the industry, the REL option would even further reduce interest in a stable, low-polluting alternative to the ICE. The legal viability of existing technical approaches would be assured, and the looser control program would quickly become established and difficult to adjust. In short, a policy of simply relaxing the emissions standards forfeits flexibility.

ALTERNATIVE TECHNOLOGY

Several of the alternatives discussed earlier offer stable, low polluting engines with no insurmountable obstacles inherent in the technology itself. (Of course it will cost time, money, and political energy to make a change, and side effects should also be considered.) The argument for changing the technical basis of the emissions control program to an advanced engine technology rests upon two propositions. The first is that the technical trade-offs now plaguing control of the ICE are not likely to be drastically eased by the discovery of some supergrade catalyst—what insiders now refer to as the "magic bullet." The second is that the nation, whether out of considered rational judgment or sheer political will, will in fact insist on the objective of 90 percent abatement now set by the Clean Air Act and

---

[10] A strong statement of this view is to be found in the *RECAT* study [Office of Science and Technology 1972].

will therefore hold to the current emissions standards. If these propositions are true, the shift to alternative technology is both necessary (to achieve the objectives) and economically wise (to avoid the high fuel, maintenance, and enforcement costs of an unstable technology).

Since data on alternative technologies are even more uncertain than those for current technology, we will not attempt to make direct estimates for the various propulsion systems discussed above. Instead, we will define an alternative technology option as one that achieves Level 0 stability for an incremental manufacturing cost (including research and development (R&D) and retooling/retraining costs) of $500 more than the ICE with 1973 controls. Since the hypothetical technology is stable, no excess maintenance cost is levied. (In fact most options would provide actual savings in maintenance and replacement.) Fuel costs are assumed to be the same as for the gasoline powered ICE (even though most of the technical alternatives run on cheaper fuels). These assumptions about operating costs are reasonable, or even pessimistic, for Rankine cycle and stratified charge engines. Thus we establish an *a fortiori* argument. If the hypothetical alternative technology compares favorably with the current control strategy even under these harsh assumptions about the manufacturing costs of an advanced engine, then the real alternative, likely to cost less, will be even more attractive.[11]

Since we are assuming that none of the alternative technologies can be prepared before the model year 1981, this option would have to include some interim program. There are two obvious possibilities. The government could hold to the established standards, force the industry to install the ICE control devices now contemplated, and simply tolerate the resulting instability during the 1975-1980 period. We have labelled this option EST/ALT. Responsible officials might choose it to maintain the political integrity of the program and/or to gain the benefits of slightly lower pollution. The EPA might even allow sales of model lines that came close to the standards but somehow failed to pass the rigid prototype test. In that case, nonprohibitive fines for noncompliance would be imposed during the interim period. This approach is described in greater detail in Chapter 4. The second possibility would be to relax the emissions standards during the interim period to reduce costs and to help shift Detroit's attention to new technologies. This option is labeled REL/ALT.

Fifteen-year program costs for the two advanced technology options differ in accordance with various cost estimates made for the interim program. For each alternative we project low, medium, and high program costs reflecting different assumptions about the interim period. These are presented in Table 3-4.

There is a formidable difficulty with the alternative technology option: many people will have to change their hearts and minds about what sort of

---

[11] The NAS panel provided an estimate of the costs of conversion to an alternative technology which under very drastic assumptions ranged from only $8 to $150 per vehicle [1973, p. 103].

Table 3-4. Low, Medium, and High Estimates of Fifteen-Year Program Costs with a Shift to Alternative Technology in 1981 at a 5 Percent Discount Rate

Cost Assumptions	Fifteen-Year Program Cost ($ Billions)	
	Established Standards 1975-80 (EST/ALT)	*Relaxed Standards 1975-80 (REL/ALT)*
Low cost	51.3	44.1
Medium cost	68.0	52.0
High cost	83.7	63.5

equipment belongs under the hood of a car and what kind of research and development the federal government should conduct. The automobile manufacturers, who are doing well with the ICE, are not likely to respond enthusiastically to a program that forces them into a major reorientation of plant capacity and that localizes more of the cost of emissions control at the manufacturing stage. To promote serious development of alternative technology by the industry, the government would undoubtedly have to contribute financially to the research and development process—as it has done in other sectors of the economy. This would require major changes in the habits and policies of critical agencies of the government—particularly the congressional subcommittees dealing with pollution issues and the Office of Management and Budget, both of which have opposed such a governmental role.

The alternative technology option, moreover, would inevitably expand the scope of the policy into the problems of managing energy resources. The established policy itself would have a major impact on oil consumption and would significantly increase our balance of payments deficit in the late 1970s and 1980s.[12] Adoption of one of the alternative technologies, however, would change the basic pattern of energy consumption. Since automotive transportation accounts for a significant portion of overall energy use, it would be unwise to embark on a shift in automotive technology without a much more thorough analysis of the energy implications than is now available.

EVALUATION OF THE OPTIONS

Figure 3-1 provides a schematic summary of the basic policy choices and the two key outcomes—cost and stability—that determine their relative attractiveness. Unfortunately, it will be five or more years before it is known for certain which

[12] At a 10 percent fuel penalty, which is in the middle range of the estimates shown in Tables 2-1 and 3-2, the addition to oil imports could amount to $1 billion or more by the late 1980s. This would require an additional 200,000-ton supertanker arriving at a United States port every other day!

branch best describes the cost of any policy decision, and it will be a decade before the stability characteristics of controlled vehicles (and therefore the ultimate pollution levels) are fully known.

In this kind of uncertain situation, one very informative approach to evaluation is to look at the *expected* outcome in terms of cost and on-the-road emissions. Estimates of costs and vehicle stability just discussed span the range of possible outcomes, and reasoned judgments can be made about how likely these results are. By combining the estimates of effects and their relative likelihood, one can develop estimates of the expected outcome for each policy; these data should prove helpful in deciding among the different options.

Two such estimates are presented below. The first is based on our own analysis of the evidence available to date and is considered a "reasonable" evaluation of the different options. Then, to test the sensitivity of the conclusions, we include estimates reflecting unabashed optimism about the future of emissions controls on the conventional ICE. The point of the exercise is to show that the conclusions are the same over a wide range of disagreement about probable costs and stability levels.

Reasonable Assumptions

Reasonable assumptions about the likelihood of various outcomes are displayed in Table 3-5. For each of the policies, two sets of assumptions must be made: What is the probability that fifteen-year program cost will be nearest our high, medium, or low estimates? And, given that the cost comes out at a particular level, what is the probability that vehicle stability will turn out to be Level 1, 2, 3, or 4? From these assumptions, and the data presented in Tables 3-1 to 3-4, the expected cost and the expected Weighted Index of Reduction can be determined. We can also calculate an indicator of the emissions during the period 1985 through 1989 in relation to 1971 levels. The results are shown in the three righthand columns of Table 3-5.

Consider the assumptions for the established policy, EST, for example. Trying to be reasonable, we assume there is only a 25 percent chance that the cost will be the high estimate in Table 3-2, and a 25 percent chance that it will be the low figure. Accordingly, there is a 50 percent chance that the medium estimate is correct. These assumptions give a present value of expected cost for the fifteen years of $61.1 billion (using a 5 percent discount rate). In addition, we assume that there is only a 10 percent chance that controlled vehicles will be as stable as Level 1, and similarly a 10 percent chance that the program will produce vehicles as poor as Level 4. The result is likely to be somewhere in the middle, so we assume a 40 percent chance of Level 2 and a 40 percent chance of Level 3.

In this particular calculation the likelihood of different stability levels is assumed to be independent of the cost of the program. As the layout of the table and of Figure 3-1 imply, more complex assumptions are possible. It may be,

Table 3-5. Expected Outcomes of Alternative Policy Options Under Reasonable Assumptions About the Relative Likelihood of Different Levels of Cost and Vehicle Stability

Policy Option	Program Cost		Probability of Stability Level for Given Cost Level				Expected Fifteen-Year Program Cost ($ Billions)	Expected Weighted Index of Reduction 1975-89	Expected Reduction in Emissions 1985-89 as Fraction of 1971 Levels
	Level	Probability of Level	1	2	3	4			
EST	Low	.25	.1	.4	.4	.1			
	Medium	.50	.1	.4	.4	.1	61.1	.39	.69
	High	.25	.1	.4	.4	.1			
EST/ENF	Low	.25	.1	.4	.4	.1			
	Medium	.50	.1	.4	.4	.1	74.2	.50	.79
	High	.25	.1	.4	.4	.1			
EST/ALT	Low	.25	.1	.4	.4	.1			
	Medium	.50	.1	.4	.4	.1	67.8	.51	.88
	High	.25	.1	.4	.4	.1			
REL	Low	.25	.5	.5	.0	.0			
	Medium	.50	.5	.5	.0	.0	32.0	.24	.55
	High	.25	.5	.5	.0	.0			
REL/ALT	Low	.25	.5	.5	.0	.0			
	Medium	.50	.5	.5	.0	.0	52.9	.44	.85
	High	.25	.5	.5	.0	.0			

for example, that if the cost turns out to be at the high end of the range, then one might expect the vehicles to be more stable than if costs are low. (One also can argue the opposite.) Based on information available to date, however, the assumptions in Table 3-5 are the most reasonable, and for policy EST they result in an expected value of .39 for the Weighted Index of Reduction.

The last column of Table 3-5 gives a rough estimate of the degree to which the current objectives of the Clean Air Act would be achieved under each of the various policy options. It gives the weighted reduction in all three pollutants, stated as a fraction of their 1971 levels (the reference point of the Act). The measure is calculated for the 1985-89 period to see the results of each option after a decade—long enough to achieve the full degree of control possible under the policy. Under the assumptions described above, the established policy would yield a 69 percent reduction in this period—about 20 percent short of the objective.

Next, the table shows the same analysis for a program with enforced maintenance, EST/ENF. The relative likelihood of different levels of cost and stability are assumed to be the same as for established policy, and once again the table shows the expected costs and effects in the righthand column. Two things are worth emphasizing about the analysis of a program with enforced maintenance. First, by assuming that the probability of different stability levels is the same as for a program without maintenance (.1, .4, .4, .1) we imply that the manufacturers' basic vehicle design is not significantly influenced by the maintenance program. (Of course, cars spend less time in the higher emitting state and therefore pollute less, as indicated by the increase of the Weighted Index of Reduction from .39 for EST to .50 for EST/ENF.) Second, by assuming a maintenance program that makes cars like new, we are *extremely* optimistic about the performance of the service sector. On balance, the second is by far the more significant bias, and, if anything, the analysis is tilted in *favor* of maintenance schemes. This should be kept in mind in assessing the results and conclusions.

Table 3-5 also presents assumptions for EST/ALT and the resulting performance indicators. Note that the probabilities shown are for the cost of controls during the 1975-80 period only, and that the cost of vehicles powered by the alternate low-polluting technology has been set at an inflated level throughout the analysis. Thus the analysis is biased purposefully *against* the options involving advanced technology; this also should be kept in mind when considering the conclusions.

Finally, the table lists calculations based on relaxed standards: the case of relaxation (REL) and the option of relaxation coupled with a shift to alternate technology (REL/ALT). The estimates for these options incorporate optimistic assumptions about gains in stability with relaxation of the emissions standards mandated for 1975 and 1976. We assume that there is no chance that stability will be as bad as Levels 3 or 4, and about a 50 percent chance that stability will be as good as Level 1.

Figure 3-4 shows the data in the last three columns of Table 3-5 in

44 Clearing the Air

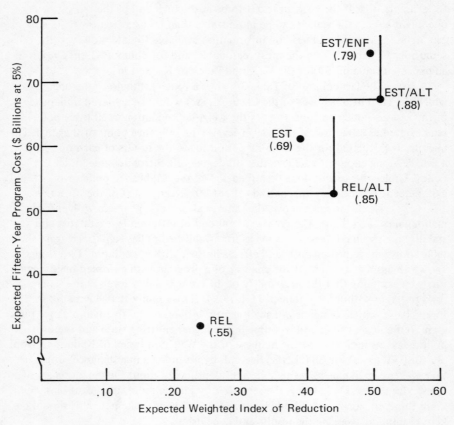

Figure 3-4. Expected Outcomes Under Alternative Policy Options for Reasonable Assumptions About Program Cost and Stability of Vehicle Emissions

graph form. Each of the five policy options is plotted, with cost on the vertical axis and emissions reduction on the horizontal. The numbers in parentheses show the emissions reductions achieved by 1985-89 as fractions of 1971 levels. The best place to be on this diagram is as close to the lower righthand corner as possible; movement in that direction means more cleanup at less cost.

The results are striking. First of all, the option of relaxing standards and switching to alternate technology clearly dominates the current policy. It provides a greater expected reduction in emissions over the 1975-89 period, and at an expected cost saving of around $8 billion in present value terms. It also more closely approaches the 90 percent objective. It is better in every dimension, despite the bias of the cost analysis *against* alternate technology.

What about enforced main.enance? As Figure 3-4 shows, the option

EST/ENF does achieve a greater emissions reduction than the established policy, at a price. But it is dominated by option EST/ALT, which has established standards for the 1970s and a shift to alternative technology in the early 1980s. EST/ALT has a higher expected Weighted Index of Reduction and an expected cost advantage of over $6 billion. The relationship between the options REL/ALT and EST/ENF, on the other hand, is not one of dominance; still, the clear preference between the two is for REL/ALT. It achieves tremendous cost savings (around $20 billion) with only a small (4 percent) loss in pollution reduction over the 1975-89 period.

The obvious conclusion is that established policy, either with or without maintenance, is dominated by options involving alternative propulsion technologies. To the extent that the framework of assumptions holds, there is a clear advantage to the economy and to the breathing public to prepare for a shift to a new technology. Currently, the federal government's annual expenditure on alternative propulsion technology is around $5 to $8 million per year. The expected cost saving, in present value terms, of a shift to a clean alternate is a thousand times that amount—and with the additional benefit of cleaner air.

The analysis also strongly indicates that large-scale systems of vehicle inspection and enforced maintenance are not a good idea, even under very optimistic assumptions about their performance. The difficulties and costs of regulating 100 million individual motorists are so great that it is evidently better to insist on a stable, clean vehicle to begin with—even at considerable cost in development and manufacturing.

What about relaxing standards to save money? The saving is great, as Figure 3-4 shows, but so is the increase in emissions—even with optimistic assumptions about stability gains as a result of loosening controls. The expected value of the Weighted Index of Reduction is only .24, and the expected total emissions by the end of the 1980s are only 55 percent below 1971 levels. The nation is unlikely to accept this result, which after fifteen to twenty years of effort would still leave a palpable air pollution problem in many areas of the country.

The choice between REL/ALT and EST/ALT is more difficult; it involves a tradeoff between cost and cleanup, and neither option dominates the other. The incremental cost of the additional reduction to be gained by the movement from REL/ALT to EST/ALT is high, as might be indicated by the slope of a line drawn between the two points. On the other hand, there is an argument in favor of EST/ALT that is not captured by this analysis. The analysis assumes that the transition to alternative technology can actually be brought about—which, of course, it can if the problem is taken as a major national priority. But if Detroit might successfully resist a proven alternative, even if it is clearly superior from a public point of view, then REL/ALT and EST/ALT differ greatly. Under the REL/ALT option, if the advanced technology is not actually marketed, then the result is that produced by relaxed standards alone. Sticking with the established standards during the interim period, while an alternative is being prepared, would be insurance against a failure in implementation at a later stage.

Optimistic Assumptions

What if one thinks that the assumptions in Table 3-5 are too conservative, that current efforts are much more likely to succeed, and that the technical problems inherent in the ICE will be solved? For the true believer, we recalculate the estimates using assumptions that should satisfy the most optimistic observer. The data shown in Table 3-5 are revised to reflect the assumption that under current policy there is a 70 percent chance that vehicle stability will be as good as Level 1 and that the chance of stability reaching Level 2 or better is around 90 percent. Furthermore, we assume that if standards are relaxed, there is a 90 percent chance of attaining Level 1 stability.

These are incredibly optimistic assumptions, yet they do not affect the conclusions we drew from the former, more reasonable, set of assumptions. Figure 3-5 shows the results. Once again the policies involving alternative technology (REL/ALT and EST/ALT) are superior to the two versions of established policy.

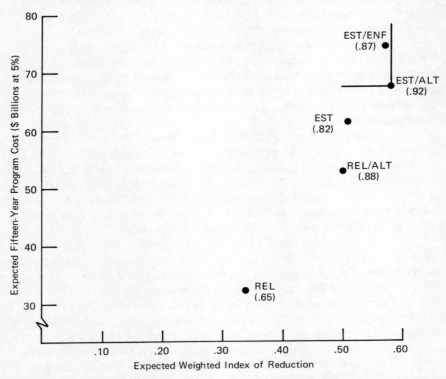

Figure 3-5. Expected Outcomes Under Alternative Policy Options for Reasonable Assumptions About Program Cost and Extremely Optimistic Assumptions About Vehicle Stability

EST/ALT dominates EST/ENF, as before. (That is, it is better in both dimensions.) REL/ALT no longer dominates established policy (EST) because the expected Weighted Index of Reduction for EST is now better. But it is only slightly better (one percentage point). Thus even with extremely optimistic assumptions about the stability of controlled ICE-powered vehicles, the advantages of a shift to alternative technology remain clear, though less dramatic than under more reasonable assumptions.[13] The policy of simply relaxing standards (REL) still is unacceptable.

On balance, the established policy appears to be on the wrong track. Even with the assumptions loaded in its favor, it is not likely to meet the objectives of the Clean Air Act and is not an efficient solution in economic terms. Moreover, rigorous implementation of the enforcement provisions now set forth in the legislation won't improve the results. The attempt to crack down on an inherently unstable technology with elaborate enforcement procedures would be an immensely expensive exercise in frustration.

If, as now seems to be the case, the government is serious enough about its objectives to contemplate gasoline rationing in Southern California and severe traffic controls in the nation's capital, then it ought to be serious enough to attend to the basic flaws in the regulatory machinery. The current standards and deadlines form a highly questionable technical approach to emissions control. Serious effort needs to be devoted to preparing a more appropriate technology.

---

[13]It is possible to make still other assumptions about stability, although it strains the imagination to consider assumptions more favorable than those reflected in Figure 3-5. One could also make alternative assumptions about the relative likelihood of different cost levels, although it would, as argued earlier, take basic revisions in the entire framework to threaten the results arrived at here. For the reader who wants to try his own estimate, all the necessary data are provided. The costs are shown in Tables 3-2, 3-3, and 3-4; the effects are recorded in Table 3-1. Following the format of Table 3-5, one can try any set of assumptions and see how the answers come out.

Chapter Four

Advanced Technology and the Problem of Implementation

Henry D. Jacoby
and
John D. Steinbruner

The preceding analysis reveals that the catalyst-controlled conventional ICE is likely to experience serious deterioration of emissions controls "on the road," and that the administrative burdens of attempts to regulate in-use vehicles are likely to be overwhelming. We argue that an advanced automotive technology with stable, low emissions ought to be developed and introduced. According to the analysis, options REL/ALT or EST/ALT are clearly preferable to the other policies as long as current objectives are maintained.

Unfortunately, such a policy runs counter to the psychological, technical, and economic momentum of the auto industry. The manufacturers, in response to the Clean Air Act Amendments of 1970, have made a strong commitment to adjustments and bolt-on additions to the conventional ICE, and in particular to catalyst-based solutions. We must ask, therefore, whether the government has the finesse and political energy to bring about a change in automotive technology and whether the cost of the shift would be acceptable. In short, can an emissions control policy based on advanced technology actually be implemented?

The problem of implementation involves more than simply achieving a technically workable automobile with an unconventional engine. The issue encompasses both the politics of concentrated private interests and the complexities of large organizations. As noted earlier, there are no insurmountable technical obstacles to the development of four or five different technical approaches, and several (for example, the stratified charge engine) are generating considerable optimism in the technical community even now. It is conceivable, however, that organizational and political barriers to their adoption might seriously jeopardize an emissions control policy based on these technologies. Thus such barriers are potential threats to the early implementation of alternative technology.

Problems of implementation are uncomfortable for the policy analyst because the topic encompasses a wealth of specific detail. Once the basic objectives and general guidelines have been decided, the details of executing a policy are almost always left to the lower levels of the bureaucracy. To penetrate to that degree of detail in advance of a policy decision is extremely difficult. Indeed many of the specifics simply cannot be known until the process of implementation has begun.[1] Nevertheless, no serious analysis can claim to be complete without some consideration of how the policy could actually be carried out. If there is any great lesson to be learned from the experience of recent years it is that the grandest of policies and the most honorable of intentions can founder on the details of everyday events.

Although it may be impossible to foresee all the complicated steps involved in carrying out a major policy decision, an examination of past experiences can provide some guidance. The history of an organization furnishes powerful clues to its likely future behavior. Specifically, the history of the automobile companies provides an important perspective on the problem of implementing an emissions control policy based on advanced automotive technology.

HISTORICAL PERSPECTIVE

The automobile industry has developed on the basis of advances in production techniques, marketing, and industrial organization.[2] The men who built the industry excelled in these areas, and the character of the industry reflects their pre-eminence. The technical and scientific aspects of the business have always been deemed important but subordinate. The industry has not invested much in basic science, and it has not developed broad technical competence equivalent to that of, say, the aircraft industry. The technical expertise of the auto manufacturers is limited to the particular product they now market. Since its inception the industry's priorities have been deeply ingrained, and they powerfully affect what it can and cannot do.

General Motors, for example, the leader of the industry, acquired its chief characteristics in the early 1920s. By then the entrepreneurial phase of the company's development was over, and control had been removed from William Durant, the man who dominated GM in the early years. It was at this time that the product line of the firm was defined—five classes of automobiles in ascending price ranges—and it was in this period that the basic form of

---

[1] The reactions of Congress are particularly troublesome in this regard. Congresssional action would be necessary to restructure the current policy if a change in its technical base were to be undertaken. Given the way in which Congress does its business, however, it is difficult to predict what problems may block passage of the appropriate legislation until a bill is actually filed. The critical members of Congress (subcommittee chairmen and other legislative leaders) are not in the habit of taking firm positions before they are confronted with an actual legislative issue and have had a chance to consider communications from constituents and other interested parties.

[2] Historical accounts of the industry can be found in *The Automobile Industry Since 1945* [White 1971], *My Years with General Motors* [Sloan 1964] and *Strategy and Structure* [Chandler 1969].

organization was worked out. Operational responsibility was located at the division level, and divisions were organized by the product they produced (Chevrolet, Buick, Cadillac, etc.). Division managers were responsible for introducing and marketing their product lines and for obtaining an appropriate return on investment. The general officers of the corporation were given basic financial control and responsibility for broad policy and strategic planning. The result was a structure in which authority over the basic activities of the company was highly decentralized; yet the men exercising that authority were closely monitored by general executives who held them accountable for the overall performance of their departments.

Naturally, the organizational form was shaped by the experience of the company during its early years, and two aspects of that experience were particularly influential. The first concerned the character of the automobile market, and the second involved particular corporate choices about automobile technology.

Very early, the general officers of General Motors began to make economic forecasts and to require that division managers do the same. In the 1921-25 period, however, they discovered that the automobile market is highly volatile and given to significant fluctuations from month to month and year to year. Economic forecasting, then as now, was not precise enough to guide efficient planning of the production process far into the future. Instead, the company worked out short-term feedback processes. Division managers were required to forecast sales and appropriate production schedules for a one to three-month period and to update these projections every ten days as data on sales and new car registrations were returned from the field. Under this procedure, the men with operational responsibility were in close contact with the day-to-day conditions of the market, and therefore worked within very short-term time perspectives.

The second formative experience involved an attempt to introduce a technical advance. Charles Kettering, generally recognized as an outstanding engineer, came to General Motors in 1919, and the company's research was organized around him. At the time he was brought in, Kettering was developing an air-cooled engine which, he argued, would have fewer parts and greater operating efficiency than the standard water-cooled engines. In the early 1920s, Kettering fought with various product divisions, trying to get them to adopt the air-cooled engine. A few air-cooled Chevrolets did reach the market, but they had such serious engineering flaws (though not all in the engine) that they had to be recalled and the project died.

In the process of pressuring the reluctant division to adopt the engine, the management neglected engineering developments on standard engines, and the collapse of Kettering's project left the company with inferior products on the market. General Motors saw this experience as a great lesson. In violation of the principle of decentralized authority, the general executives had tried to force a division to accept a technical innovation. Their failure confirmed the principle and resulted in the return of control over marketed products to the division level.

Kettering's research laboratories went on to develop many important automotive components but never again tried to introduce a completely new product embodying major technical changes.

This early experience set the organization on a course that it continues to follow today. There is great emphasis on sales and, given the need to turn a profit, on the control of costs. The time focus is basically short-term. The longest time frame is the two- to three-year period over which a model change is planned. Technology changes in this context have been the result of a long sequence of incremental improvements, all of them based on the Otto cycle internal combustion process. Usually such changes involve a small number of components, and rarely are extensive design changes in the engine, transmission, front and rear suspensions, etc., made all in the same year. Moreover, technical modifications are introduced gradually—often beginning with a single model line and spreading to other models over a period of years. More substantial changes involving a fully integrated automotive design would be incompatible with the short time perspective and division-level responsibility.

The industry does engage in some broad technical investigation. The GM technical staff—significantly located in the general offices of the corporation—has in the past explored technical alternatives to the ICE. Given the character of the corporate organization, however, such research is not a main concern of the business. Moreover, until the recent interest in the Wankel, the only alternative that the technical staff had considered in depth and with some enthusiasm was the gas turbine. The research laboratory, like other divisions of the corporation, has focused on the things the market has tended to value—speed, acceleration, fuel economy, and smooth and reliable performance. It has never been concerned with developing advanced technologies for the purpose of controlling emissions; thus the more promising of the several alternatives discussed in Chapter 2 have not received much attention. The technical staff generally has been content to monitor developments outside the industry in order to ascertain whether other firms are threatening to introduce a competitive technology.

Not all the members of the industry are exactly like General Motors, of course. Ford is more venturesome in technical innovations and is less dominated by cost considerations and immediate market conditions; it is also less efficient. The differences are relatively small, however, and too refined for the level of analysis to which we are limited here. The basic facts are that the entire automobile industry has intensely concentrated on internal combustion technology, has never experimented very seriously with alternative technical configurations, and has not had a broad base of trained scientific personnel.

This posture has worked well for the industry. Its product has become an essential element of the American life-style. Despite deficiencies in vehicle safety and durability, consumers have been willing to buy as many cars—and buy them as often—as their incomes have allowed, and stable long-term demand (though not necessarily growth) seems assured. Since the new-car market is largely a replacement market, however, consumers can and do effect short-term changes in demand by postponing replacement decisions. As a result, there is little incentive under current conditions for fundamental changes in the product line. The popularity of the ICE-driven automobile makes the long-term benefits of a technical change questionable, and variations in demand make it risky in the short run. Indeed, since several of the alternate technologies may prove more durable than the conventional ICE, they threaten the critical replacement rate.

The problem of policy implementation is also affected by the competitive situation within the industry. With a small number of large firms, the market resembles the classic oligopoly, and there is a strong presumption that some limitations on competition are tacitly arranged by the manufacturers. General Motors, with 50 percent of the market, is the dominant firm in the industry, and by most estimates GM could expand its market share substantially should it choose to do so. The reason it does not do so is the threat of antitrust action by the federal government—a threat which has become a permanent feature of the automobile business. Since no other company can beat GM in either cost or quality, and since GM cannot afford to drive the others out, the whole industry has a protective, status quo psychology which does not encourage major new departures. Competition focuses largely on subtleties such as style, image, warranty provisions, etc., which are unlikely to produce sudden and substantial shifts in the market.

Given the historical posture of the industry, it is not surprising that the firms have reacted to pressure for emissions control with ad hoc adjustments to the ICE rather than by fundamental technical change. During the 1960s, Senate hearings on Rankine cycle and electric drive engines [U.S. Senate, Committee on Commerce and the Subcommittee on Air and Water Pollution of the Committee on Public Works 1967 and 1968] generated pressure for further development of these technologies. Manufacturers responded with centrally controlled technical efforts seemingly designed to prove the impracticality of these alternatives. Obviously unmarketable prototypes were displayed and depicted as the best that could be produced. The technical efforts never reached the product divisions, and the research teams were disbanded after the hearings were held. Though few outside the industry would condone this behavior, it cannot simply be ascribed to greed, bad faith, or other base motives. Because of the way the industry has evolved, there is an inevitable difficulty in achieving the technical flexibility that the problem of emissions control requires.

One notable exception must be acknowledged, however. During World War II, the automobile industry unquestionably displayed a great deal of technical flexibility in shifting to war production [Lowenthal 1971]. Overall, the industry produced one-fifth of all war materiel in the United States—from machine guns to aircraft[3]—and the process of conversion was exceedingly rapid.

---

[3]According to Donald M. Nelson [1946], during the war the auto industry produced 75 percent of all aircraft engines, 33 percent of all machine guns, 80 percent of all tanks and tank parts, 50 percent of all diesel engines, and 99 percent of all Army motorized units.

Chrysler Corporation, which had never produced a tank before, picked up design blueprints in June of 1940, broke ground for a plant in September of 1940, and began regular production in April of 1941. General Motors selected a site for a machine gun factory in November of 1940 and began production in April of 1941. Like all the auto manufacturers, Ford had been kept ignorant of air transport technology by a jealous aircraft industry until late in 1940. In February of 1941, the government decided to have Ford produce the B-24 and eighteen months later, in September of 1942, planes began rolling off Ford assembly lines. In a similar effort, General Motors produced over 13,000 torpedo bombers and fighters during the four years of the war. All this was accomplished while the industry was losing nearly 400,000 workers to the draft (two-thirds of the pre-war labor force) and was being forced to train some two million workers. Under the right conditions, then, the automobile industry *can* make radical changes in its product line.

If the wartime performance of the industry was impressive, however, so were the conditions under which it was achieved. All automobile production was stopped. The government, in addition to providing a guaranteed market, supplied $1.5 billion in capital investment (half of the pre-war capital assets of the industry). Normal rules for profit and return on investment were relaxed as a contribution to the war effort, and the entire undertaking was in the context of an overwhelming national commitment. The war experience shows us what is possible, but it does not tell us what to expect with business as usual.

This discussion puts the problem of implementation in focus, however. A major technical shift in the automobile industry would require a major adjustment in its normal procedures—procedures rooted in historical experience and in basic market forces. The industry will resist these changes with both organizational inertia and conscious opposition. Nonetheless, the World War II experience illustrates that a substantial technical change is not impossible.

A STRATEGY FOR IMPLEMENTATION

Based on the foregoing discussion, we can identify a number of requirements for successful implementation of an advanced technology.

First, *the designs and production techniques must be developed by the established manufacturers.* Occasionally it is suggested that an outside entrepreneur might be induced to enter the automobile industry with an advanced, low-polluting technology—thus forcing the established but recalcitrant companies to follow suit. Unfortunately, this is not very realistic. Entering the industry would cost about $1 billion—just to produce and distribute automobiles— and even then a new entrant would have little hope of being competitive [White 1971, pp. 54-76]. Control of a marketing, dealership, and service network is critical to successful operation, and a technical entrepreneur is unlikely to be able to compete in these areas.

No doubt suppliers such as Bendix, DuPont, Ethyl Corporation, Mobil Oil, etc., can and do contribute new technical developments. And a great deal of pressure can be (and is being!) generated by innovative foreign manufacturers such as Honda, Mazda, and Mercedes-Benz. But since none of these firms on the fringes of the market can take over the production of 10 million vehicles a year for American motorists, a successful policy must bring about a technical change by those firms that will continue to dominate the market.

Second, *the impetus for change must penetrate to relatively low levels of the manufacturing organizations.* There is no point in having the central technical staffs of the companies develop the desirable technologies if the marketing, production, and basic engineering departments of the product divisions are not engaged. A successful strategy, therefore, must allow enough time for the product divisions to develop the basic technology and to work it into a marketable product.

Third, *the shift to an alternative nonpolluting technology, if it is to be timely, will require constructive direction and participation by the government and strong incentives for the manufacturers.* The suggestion has been made from time to time that the Congress simply legislate a ban of the internal combustion engine. In the current context, that would be like erecting a barricade in front of a fast-moving train and doing nothing else. It is unlikely that a barrier could be made strong enough, and even if it could the results would be intolerably messy. As noted earlier, over 10 percent of United States GNP is tied up in the internal combustion engine. In matters of such scale, someone must carefully chart the new direction and lay the tracks. If the government is going to intervene in the automotive market on behalf of the public interest, it must participate in those tasks or risk a great waste of resources and loss of opportunities for environmental protection.

And fourth, *the time frame of policy must be greater than that allowed by the current deadlines, even assuming one-year extensions of both.* EPA's extension of the 1975 and 1976 standards by one year each has not shifted the technical focus of the industry. The manufacturers still must prepare the catalyst-based technology, and that will continue to absorb their energies. The extension decisions will not seriously deflect the highly unfortunate momentum of established policy, and another intervention will be required.

If the resolve to pursue the established objective remains, then there should be an immediate government initiative to remove the substantial barriers to advanced technology now inherent in the situation and to promote its development actively. A natural occasion is provided by the publication of the National Academy of Sciences study and by the ample testimony on the problems of current technology provided by the companies in support of their requests for extension.

An initiative flowing from these principles can be formed from two components: a revised structure of standards and fines and new incentives for research and development.

New Structure of Fines

Section 205 of the Clean Air Act provides a fine of $10,000 per vehicle for any cars put on the market that violate the standards. This provision should be revised to establish a two-part structure of fines for the 1975 through the 1980 model years. A relaxed interim standard should be established, and a lesser fine (i.e., one that would not effectively prohibit sale) should be levied on cars that fall in the zone between this interim standard and the full 1975 and 1976 goals.[4] Such a change would include the following revisions of the Clean Air Act:

1. Instruct the Environmental Protection Agency (EPA) administrator to establish an interim standard for the 1975-80 period. It should be set at the tightest levels considered technically "reasonable." The emissions rates set under option REL discussed above give a rough idea of what the limits might be.
2. The $10,000 per car fine for violation will apply to the interim standard for the period of its life, 1975 through 1980. In 1981 the interim standard will expire, and the prohibitive fine once again will apply to the standards now set for 1975 and 1976 model years. Certification of vehicles under this provision will take place at the prototype stage, as it does now.
3. Models meeting the interim standard but failing to achieve the 90 percent reduction mandated in the Act will also be considered in violation. But they will be subjected to a per-vehicle fine that can actually be collected, and that will provide both an incentive to action by the manufacturers and a source of revenue to be used for perfecting a socially desirable alternative technology.
4. The level of the fine for violation of the interim standards would be set in the amendments to the legislation, presumably with some discretion left to the EPA administrator to deal with unexpected circumstances, hardship cases, and the like. The fine could be set in a number of ways: uniform for all cars, or based on weight, horsepower, value, or some other criterion. The best scheme would be a fine stated as a percentage of the sticker price of the vehicle, excluding accessories. We recommend a fine of not less than 5 percent nor more than 10 percent of the vehicle cost. For each vehicle type and engine class, the fine would be imposed on the basis of the prototype test result.
5. The amendments should require that the EPA administrator establish a system of auditing the quality of the cars actually coming off the assembly line. The full federal test used in prototype certification should be applied. Following well-established statistical procedures used in the field of quality control, only a fraction of the vehicles produced need be tested. It also should be possible to design a quality audit system that could satisfactorily handle the special difficulties of measurements on brand-new engines (the so-called "green engine" problem) so that the expense of mileage accumulation can be avoided.
6. The quality audit provides a second chance for a vehicle which failed the proto-

---

[4] A similar set of financial incentives have been recommended by D. N. Dewees [1971].

type test to show compliance with the standards. This will provide an incentive for manufacturers to continue to seek improvements as the model run proceeds. In the event the quality audit shows that a vehicle meets the 1975-76 standards, from that point on the fine would be lifted from that particular model and engine type.

This form of quality audit requires that the average emissions results at the assembly line be accepted as the measure of performance for different model types. And to attain a large sample (and thereby reduce testing costs), the audit results should be aggregated for all assembly plants of a particular manufacturer producing a particular vehicle type. (The regulations also might allow a manufacturer to update the deterioration factor computed at the time of the prototype test.)

7. Given the importance attached to the quality audit under this procedure, it may be argued that the prototype test should be dropped altogether. The reason for retaining it is that it serves to reduce uncertainty for the manufacturers. Once a vehicle has been certified, the manufacturer can proceed with confidence that the model run can be sold. Without the test, automakers might face an unexpected shutdown or additional expense right in the middle of the model run.

In the event the quality audit shows an emissions performance *worse* than that exhibited in the prototype test, the EPA administrator should have the discretion to impose whichever of the two fines may be appropriate. The Act should leave the administrator the flexibility to handle such a decision in a reasonable manner. Manufacturers found in violation should be allowed time to correct the problem or to reorient their production plans without unacceptable economic waste or social disruption. (On the other hand, good performance by one model type should not be allowed to offset violation by another—in effect averaging over *all* vehicles. If this were permitted, important incentive effects of the policy would be lost.)

These seven recommendations constitute a coherent and reasonable way to adjust the current policy to avoid unnecessary waste and promote quick adoption of new developments in technology. Details remain to be worked out (e.g., the fine could be placed on the manufacturer or on the car itself), but the overall time scheme is feasible and is in the spirit of the 1970 amendments to the Act.

Incentives for Research on Alternative Technology

A critical component of the strategy described above is that a technology capable of meeting the 1975 and 1976 standards actually be made available by 1980 when the interim standards would expire. In this context, "available" means that the propulsion system is fully developed and proven under conditions of mass production. It has already been observed that the R&D effort by domestic manufacturers has not been effective; the success of small-scale foreign manufacturers in achieving clean designs reinforces this argument. Therefore, in order to reallocate resources to serve broader social needs, the government needs to play a large role in the development effort. To achieve this, the following changes are recommended:

1. All fines collected under the scheme just outlined would be earmarked for an Automotive Technical Development Fund. In fiscal years 1974 through 1976, the fund would be supplemented with Congressional appropriations of $50 to $100 million per year. After 1976, the fund would receive the fines only; thus the whole effort (which is not unprecedented but is unusual in the history of U.S. government-industry relations) would phase out automatically as the problem is solved. When the cars meet the standards, there will be no more money collected from the manufacturers (and ultimately from motorists) for technical development.
2. The fund would absorb the current Advanced Automotive Power Emissions (AAPS) program within the EPA, and would be administered either by a new agency controlling research on energy problems or by the National Science Foundation.
3. The money would be used to contract research by the automobile manufacturers and others on new propulsion technology and on solutions to production problems that may arise. The program would have the following development goals: (a) engines with low emissions; (b) engines that require no maintenance and no cooperation from the owner in order to maintain low emissions on the road; (c) engines that have low fuel consumption; (d) advances in vehicle design that will preserve or increase safety while achieving the other goals; and (e) demonstration that designs are suitable for mass production.

The model for this kind of effort is the Programmed Combustion (PROCO) program carried out by the Army Tank Command. The Army had a long-range military need for a low-emissions vehicle and hired the Ford Motor Company to do the work. Similarly, since the long-run needs of the environment and the energy economy do not necessarily correspond to the needs and the operating procedures of particular corporations, government guidance (and therefore financing) is needed. As argued earlier, the long-run stakes in terms of the environment, the economy, and the balance of payments are simply too great to allow this opportunity to pass.

Feasibility of the Proposal

If such two-part change in policy could be set up, it might solve the organizational problems of implementing a technical change. It would allow a more realistic planning horizon than the 1975 and 1976 deadlines. It would set up governmental machinery with a clear mandate to develop specified technologies and with staff to keep track of the vast amount of detail involved in conversion. With $200 to $500 million per year available in the Technical Development Fund, the government would have the financial capacity to influence the manufacturers' research and development efforts, and it would have resources to ease the transition for the labor force and the service industry. The fund also would encourage serious policy studies of alternative technologies that meet basic energy requirements as well as contribute to pollution control.

The major question is whether anything like the arrangement just outlined could be achieved politically. Since it would require new legislation, the policy change could be blocked at a number of points in the legislative process. The automobile manufacutrers would not greet the plan with enthusiasm. Not only would they oppose the imposition of a new technology, but they would fight the general principle of government regulation. A program of the sort envisaged would put the industry under a much stricter set of regulations than it has ever experienced, and the companies would very likely resist the precedent. Whether or not they would be pressured sufficiently by their problems with the 1976 standards to acquiesce to such an arrangement is a major question.

No matter how the industry reacted, opposition to the scheme might well come from those who control tax policy—the Treasury Department, the Office of Management and Budget, and the House Ways and Means Committee. These segments of the government are not particularly concerned with pollution control but are concerned with taxing arrangements. They are likely to interpret any fine imposed on vehicles emitting in excess of the standard as a tax in effect, and they tend to oppose special purpose taxes. Their opposition might be diminished by a provision that terminated the Technical Development Fund after, say, seven years; whether such a provision would induce them to approve the proposed arrangement, however, remains a major question.

Another question is which agency should administer the technical development effort. EPA is the logical choice and would certainly claim jurisdiction. Congress, however, does not seem to trust the technical competence of that agency and may be reluctant to give it the mandate, incentives, and monitoring authority implicit in the technical development fund plan. That reluctance would be reinforced by inevitable disagreements within the technical community about the desirability of various alternative technologies. The advantages of those technologies, it must be recalled, have a great deal to do with factors that transcend strictly technical questions, and the established authorities on automotive technology are not always well informed about these broader issues. Congressional reluctance might be overcome by a properly chosen advisory board to the Technical Development Fund, but that again is a major question.

In announcing on June 29, 1973, a series of actions regarding energy problems, the President suggested the creation of an Energy Research and Development Administration. Such an agency would be a possible locus for the technical development effort since problems of gasoline consumption are so integrally related to the technical problems of emissions control. If that agency is not created or if its mandate cannot be extended to the emissions control aspects of automotive technology, then we suggest that administration of the technical development effort be entrusted to the National Science Foundation.

Finally, other agencies of the government would find their interests involved in such a proposal because of its potential side effects. The Department of Transportation also would claim jurisdiction over the technical development effort and would resist invasion of its territory in the name of pollution control. The antitrust division of the Justice Department might well seek to impose a number of constraints in order to preserve competition within the industry. Just getting the various components of the executive branch to agree would be a hazardous undertaking.

Without the test of an actual proposal, it is difficult to determine the strength of these political barriers and the cost of overcoming them. Intuitively, however, we consider these countervailing forces strong enough that the Administration would have to make a major commitment to this strategy and the President would have to give the project his backing in order for there to be any chance of success. The power of the White House would certainly be required to bring the bureaucracy into line and would very likely be required to force congressional action.

Why then might the President undertake such a commitment, given its inherent difficulty and the multitude of competing claims which he will have to face? Apart from the motivations of the particular person involved and his peculiar political situation, where are two possible reasons, one a threat and one an opportunity.

The threat is entailed simply in the fact that the President will inevitably be involved if the situation evolves into a dramatic confrontation between EPA and the industry. If the vehicles cannot pass the certification tests, the President will have to contemplate the consequences of disrupting automotive production. If, in addition, the environmentalist movement retains the political force it has displayed over recent years, then the Administration would feel uncomfortable being caught between it and the industry. The issue of pollution control has assumed a moral character that goes well beyond what can be captured in economic damage calculations. The clash between the moral impetus of environmentalist forces and the economic inventives of the industry might be quite serious if the situation is not managed well. The President may want to take strong preventive action.

The aspect of opportunity concerns the broader issue of technology in American society. A broad consensus of American opinion has come to recognize that defense and aerospace applications have dominated American technical development far too much. The emphasis in these areas has not been in balance with the basic needs of the society, and in the long run this hurts both the scientific community and the society as a whole. The scientific community suffers when its efforts cannot be sustained, as is inevitably the case when such imbalances occur. The society is hurt because areas of critical need remain underdeveloped. The automobile industry is a perfect embodiment of the problem. Its technical development has been far too sluggish, given the importance of its product in the daily lives of the citizenry. With the exception of World War II, its concerns have been largely remote from the realm of defense, and hence the government has not provided it with the technical stimulation given the aircraft and electronics industries. It is time that priorities change, and the question of emissions control opens the way to begin to effect this change. The significance of this opportunity lies far beyond air pollution, and this consideration might be attractive to a President who has vision.

Chapter Five

Emissions Measurement and the Testing of New Vehicles

Milton C. Weinstein
and
Ian D. Clark

The 1970 Amendments to the Clean Air Act establish emissions standards for all vehicles sold in the United States for the three principal automobile pollutants: hydrocarbons (HC), carbon monoxide (CO), and nitrogen oxides (NO_x). These standards apply not just to vehicles when new but throughout their "useful life." Complicating the matter, but at the same time affording the Environmental Protection Agency (EPA) a measure of flexibility, is the fact that the legislation does not specify precisely how the emissions standards are to be enforced. For example, it is unclear whether the standards are to apply to each vehicle individually, or to an average of all vehicles, or to an average of vehicles within each engine classification. A standard applied to average emissions of all vehicles or all vehicles of a certain type is less stringent than the same standard applied to each vehicle individually.

The selection of the testing procedures to be used also affects the stringency of the de facto standards faced by the manufacturers. The choice of tests and the decision whether to enforce the standards on each vehicle or on average emissions are important elements of the actual *definition* of the standards. The tests and enforcement standards, in turn, will affect the production decisions of the manufacturers and the actual emissions from their products.

The 1970 legislation alludes to three components of an enforcement program. The first is the testing of prototype vehicles before mass production has begun. This is the only part of the testing package that has actually been implemented so far. The second is the testing of vehicles at the assembly line. And the final element is state testing of vehicles "on the road," with the associated warranty and recall procedures. The selection of the tests to be used, the criteria to be applied (averaging or not), and the number of vehicles to be tested are left to the administrator of the EPA.[1]

It is essential that the EPA formulate an effective, equitable, consistent, and comprehensive vehicle testing strategy. Decisions about each of the three stages of testing (prototype, assembly-line, on-the-road) cannot be made independently of each other, nor should they be made without considering the enormous importance of such decisions in carrying out the intent of Congress in the 1970 legislation.

FRAMEWORK FOR EPA DECISION-MAKING ON TESTING

In terms of actual emissions and the magnitude and distribution of costs, the effect of the testing policy set by the EPA will depend on the responses (and counter-responses) of state agencies, auto manufacturers, and car owners. Figure 5-1 summarizes the essential elements of the framework in which the administrator of the EPA must make decisions about testing of prototypes, new vehicles, and in-use vehicles.

The Environmental Protection Agency

The principal decision-making unit is, of course, the EPA, shown in the box at the top of the diagram. The administrator of the EPA is authorized under Section 206 of the Clean Air Act (CAA) to establish procedures, rules, and criteria for the testing of prototype vehicles and for the testing of cars at the assembly line. He must establish procedures for simulating driving conditions, select instruments for measuring emissions, and determine standards by which to evaluate these measurements. In order to regulate on-the-road emissions, the administrator has three additional policy instruments. First, under Section 110, he is empowered to determine whether or not each state should require in-use inspection in order to meet the national ambient standards. Second, under Section 207 (c), he is authorized to initiate a recall of any model vehicle of which a "substantial number" are in violation of emissions standards. And third, under Section 207 (b), he must determine whether each state's inspection procedure is sufficiently correlated with the federal certification test to be eligible for inclusion under the federal warranty (which transfers all costs of failing of such a test to the manufacturer, provided that the car has been properly maintained).

---

[1] For the reader who finds he wants more background material on the automobile testing provisions of the 1970 Clean Air Act Amendments, the steps that have been taken by the EPA and the states in response to that legislation, and the basic characteristics of the principal emissions tests, the National Academy of Sciences report [1973] is an excellent reference.

The Automobile Manufacturers

On the basis of the EPA's decisions in these areas, automobile manufacturers (shown in the figure in the box on the left) will design and produce motor vehicles, reacting to the testing and warranty requirements.

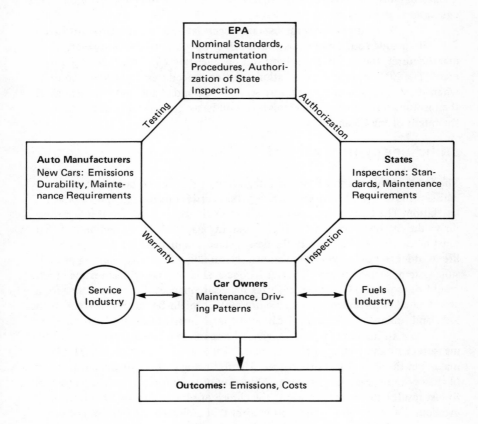

Figure 5-1. Framework for Decision-Making on Testing

The States

On the basis of the criteria for authorization under warranty and recall, and on the basis of the EPA's stipulations for implmentation plans that will meet the ambient air standards of the Clean Air Act, the states (shown on the right) may produce vehicle inspection systems featuring various testing procedures and criteria. The EPA will then be in a position to authorize or not authorize such inspection systems for use under warranty or recall provisions of the Act.

Car Owners

Owners, subject to whatever state inspection there may be and protected by whatever warranty may be in effect, will make decisions about the maintenance they will perform.[2] We shall argue later that the frequency and quality of emissions control maintenance will be crucial in determining the actual emissions of vehicles on the road.

The outcomes of the overall process, shown at the bottom of Figure 5-1, will include actual emissions and a variety of costs—to the car owner, the manufacturer, the dealer, the service station, the states, and the federal government. The EPA must set its regulations, testing procedures, and criteria so that when states, manufacturers, and owners have responded and counter-responded, the resulting costs and residual emissions will be acceptable and consistent with the intent of the Clean Air Act.

STRUCTURE OF THE ANALYSIS

The focus of this chapter is on the integration of the various components of emissions testing into an overall strategy that carries out the mandate of the 1970 legislation. The basis of the discussion is a vehicle technology that is inherently dirty: the catalyst-controlled ICE discussed in Chapter 2. It is a technology that must be carefully monitored in the design stage and maintained during its road life in order to hold down its emissions. (Naturally, with a nonpolluting propulsion system, such as electric drive, it is conceivable that no vehicle testing at all would be required.) Chapter 6 analyzes the consequences if the conditions for a coordinated enforcement system are not met, as might be the case if some of the links and functions in Figure 5-1 cannot be implemented effectively in practice.

In the next section is a general discussion of the issues of emissions measurement and testing, including many of the details of recent federal rulemaking in the area of test procedures. The following section introduces some of the concepts used in subsequent discussion of the appropriate criteria to apply to test results, the use of averaging, the choice of sample size, and the critical problem of deterioration. The section after that addresses the question of prototype testing. Decisions on this subject have already been made and are published in the *Federal Register 36* (April 30) [1971]. Most of these rules could be changed, however, and careful consideration should be given to each of them in the context of the total testing strategy ultimately chosen. Some of the issues covered are the following:

What test should be used? The federal Constant Volume Sampling (CVS) procedure has been designed to test for—indeed, to define—"true" emissions; it forms the basis upon which the rest of the testing policy must be built. Is it adequate?

---

[2]Interacting with car owners and affecting their driving and maintenance decisions are two important sectors: the service industry and the fuel industry.

Should the standard apply to average or individual vehicles? Currently, all prototype vehicles tested must meet the established standard, although liberal allowances for retesting exist; it may be desirable to apply the standard to the average of the vehicles tested and to impose more restrictions on retesting.

How many vehicles should be tested? The number of vehicles tested is critical in determining what standard is really being enforced. Also, the costs of testing strongly influence this decision.

How should the nominal test standards be set? The numerical standards used in evaluating test results depend on what kind of test is used, whether hot or cold starts are used, and what the other test conditions are. They also should depend on whether averaging is allowed, how many vehicles are tested, and whether retesting is permitted. In short, for a fixed target level of emissions, the nominal standard should depend not only on the test procedure but also on the evaluating criteria.

How should 4,000-mile and 50,000-mile data be used? Currently, data from vehicles driven 4,000 miles are extrapolated to 50,000 miles using durability data from a small sample of vehicles tested up to 50,000 miles. How is this best done, how many of each type of vehicle should be used, and how should the results be interpreted?

What maintenance should be allowed on durability vehicles? The advent of catalytic converters by 1975 will make this decision critical, since catalysts generally do not last up to 50,000 miles.

How should mileage on durability vehicles be accumulated? To what extent should this mileage accumulation procedure simulate actual driving conditions?

Is prototype testing needed? It has been argued that perhaps there is no need to test prototypes; assembly-line testing alone could be used in combination with a strict on-the-road testing program to enforce the emissions standards. Of course, this may be a moot point since Congress has stipulated that the EPA *must* require retesting of prototypes. To what extent, however, should prototype testing be relied upon for enforcement?

In the subsequent section the closely related problem of assembly-line testing is discussed and a similar set of questions must be dealt with:

What test should be used? A number of tests have been suggested. Among them are the federal CVS Cycle, the federal "Short " Cycle, the Seven-Mode Hot Cycle, and the simple Idle Test. The decision must be made in view of the correlation of these tests with the CVS Test itself. A further possibility would be to supplement a short assembly-line test by a quality audit using the long CVS Test.

Averaging or not? This question is expecially critical if 100 percent assembly-line testing is considered.

How many vehicles should be tested? The same considerations apply here as for prototype testing.

How should the nominal test standards be set? Should the durability results from prototype tests be used to extrapolate from observed assembly-line emissions, so that the extrapolated emissions can be required to meet the "useful life" standards at 50,000 miles? Should an emissions level be set which may be exceeded by only 1 percent, or 5 percent, or 0 percent of all vehicles? In addition to these issues, the same considerations apply here as for prototype testing; that is, the nominal standard need not be the same as the target level of emissions.

Is assembly-line testing needed? Perhaps prototype testing combined with state inspections, an effective warranty, and the threat of recall form a sufficiently comprehensive enforcement program.

The last section of this chapter and Chapter 6 continue the analysis of testing and enforcement to the third stage in the process: the testing of on-the-road vehicles by the states. Once again, a complex set of issues remains to be settled:

What should be the criterion for approval of state inspection programs under the warranty or recall systems? What type(s) of test(s) should be allowed? What pass-fail criteria are equitable to both car owners and manufacturers? Are these measures cost-effective?

What maintenance should manufacturers be permitted to require under the warranty? How should this correspond to mileage accumulation under prototype testing?

What should be the test(s) and criteria used in recall? Questions about averaging, sample size, nominal standards, and correlation are all involved here.

Is a warranty needed? In making this decision, it must first be decided whether a warranty can be implemented and, if so, whether it can be effective.

Should all states be treated the same? Whether or not a state must implement an inspection program depends on its ability to meet the Clean Air Act ambient standards without such a program. Can different test regimes be satisfactory for different states? Should different parts of states (urban and rural) be treated differently?

MEASURES AND MEASUREMENT: WHAT TO MEASURE AND HOW TO DO IT

The difficulty of emissions measurement and enforcement centers on the relation (or lack of one) between instrument readings during a test and the pollutants emitted during actual vehicle use. This problem of quantification arises in many of the EPA's pollution control activities, and the structure developed here is applicable to other pollution measurement problems. But it is in emissions from mobile sources—where the emitters are great in number, operate under a wide variety of conditions, and emit pollutants that are difficult to quantify—that measurement presents the most challenging problems for the EPA.

There are four basic measurement-related problems that contribute to the difficulty of interpreting and enforcing the mobile source provisions of the Clean Air Act. They are the problems of (1) precision of measurement, (2) sample variance, (3) test mode comparison, and (4) test-to-road correspondence. Since they pervade the subsequent discussion, it is useful to explore each in some detail.

The *precision-of-measurement* problem occurs because vehicles tested and retested using a particular procedure (such as the federal CVS Test, the Key Mode Test, or the Idle Test) typically will not yield the same emissions scores every time. The reason for this is not that the emission characteristics of the vehicle change between tests, but rather that the test measurement is subject to error. The error usually consists of three components. The first is the actual measurement error of the detection instrument being used (e.g., flame ionization detector, nondispersive infrared detector). Though the magnitude of this error is small for the instruments currently used in emissions testing, it will be significant at the proposed 1975-76 levels. The second component of error results from the handling of the exhaust prior to measurement, and includes errors in dilution with air, variation in the background concentrations of pollutants in the dilution mixture, system leakage, and reactions of exhaust pollutants prior to measurement. These factors are especially important under the federal CVS procedure. The third component of the error occurs under test procedures that run vehicles through various operating cycles (from one as simple as a two-minute idle to a cycle as complex as the twenty-three-minute CVS procedure). These cycles are not exactly reproducible; due to human and mechanical errors, even dynamometer-driven cycles are subject to variation from test to test, resulting in variation in observed emissions.

The precision-of-measurement problem aside, when nominally identical vehicles (for example, those just off a single model production line) take a nominally identical test, the results differ. This may be referred to as the *sample variance* problem. The differences are due not to variation among test measurements (measurement variance) but to variation among vehicles (production variance). The decision on whether any group of vehicles (such as a whole model line) complies with the emissions standards therefore depends on the sampling and test averaging procedure. The very fact that variance exists means that there is a finite chance that some vehicles will fail the test no matter how low the average vehicle emissions are. The resultant "effective standard," to which the vehicle owner or manufacturer will react, will differ from the nominal standard by an amount that depends on the sampling procedure. This concept is discussed in greater detail in the section on "Concepts Used in the Analysis."

Next, there is the *test mode comparison* problem. No emissions test measures all the relevant polluting characteristics of an automobile. Different test procedures measure different characteristics[3] and combine them differently

---

[3]An example of a polluting characteristic is the rate of CO emissions from an idling engine ten seconds after warm-up.

into the final test score. It is therefore very difficult to compare the results of tests with varying modes. A particular car may score higher than another under one test and lower under a different test.

Finally, there is the problem of *test-to-road correspondence*. A less-than-perfect test can neither accurately reproduce the relevant vehicle operating conditions nor accurately measure all the relevant emissions. There will always be some difference between the emissions that are implied by the test score for a particular vehicle and the actual on-the-road emissions that are produced by the vehicle.[4] The relationship between the two depends upon specification of the relevant aggregates of on-the-road operating conditions and emissions, and upon the test rule-making, including choice of instruments, sample selection, and setting of numerical standards.

Any effort by the EPA to enforce the Clean Air Act emissions standards faces these four problems. The test-to-road correspondence problem arises as soon as we abandon hope of monitoring *all* vehicles *all* of the time and rely instead on a "summary" driving cycle, such as the twenty-three-minute federal cycle and the 50,000-mile, mileage-accumulation schedule. For simpler test schedules, such as the three Key Mode steady states or the pure Idle state, the test-to-road correspondence naturally is worse. This is not to say that the simpler tests are useless. On the contrary, considering their low cost, it may be desirable to sacrifice realistic test-to-road correspondence in favor of simplicity and economy, provided that the test scores are appropriately extrapolated to the best available estimates of corresponding on-the-road emissions.

As soon as vehicles are subjected to at least two different tests, the test mode comparison problem arises. If the EPA administrator chooses to require assembly-line testing using, say, an Idle Test, and if he decides to authorize state inspections using the Key Mode Test, then the scores of these tests will have to be adjusted to correspond to the federal CVS Test. The methods of dealing with this problem are discussed in the sections in this chapter on "Concepts Used in the Analysis" and "Prototype Testing."

The sample variance problem arises because it is generally possible to test only a sample of vehicles rather than the entire population. The problem becomes even more serious if a test is to be used to pass or fail an individual vehicle, as state inspections may do. The emissions of a vehicle may be well within the normal range for a particular line of cars yet exceed a standard set at the average level for that line of vehicles. Indeed, it would not be surprising to find that half of those vehicles would fail such a test. These issues are discussed at length in the section on concepts.

Finally, the precision-of-measurement problem is always with us. Until a perfect instrument is found, measurement error will always be present, although instrumentation research may improve precision.

---

[4]The test-to-road correspondence problem is given special attention in the National Academy of Sciences report [1973].

In the context of this complex set of measurement difficulties, the EPA must define vehicle emissions standards and develop procedures for enforcing them. A coherent testing strategy begins with some suitable method of aggregating the complex set of phenomena that constitute on-the-road conditions. Further, the pollution measurement itself must be defined and appropriate instruments selected. Finally, given the judgments made about aggregation and measurement, there are many subsidiary choices, including the design of sampling procedures and the determination of nominal standards for enforcement. It takes all these components to complete a testing and enforcement strategy. We will look into the aggregation and test scoring problems in some detail and then turn to the specific standard-setting and enforcement choices that are presented to the EPA by the Clean Air Act.

Definition of Emissions Standards: Specification of a Method of Aggregation

The rate of emissions of any pollutant (using a measure of mass per unit time or distance) from any particular vehicle depends on when, where, and how the vehicle is operated. The Clean Air Act specifies a 90 percent reduction in emissions; the question is, emissions under *what conditions?* How should the driving conditions be aggregated? The question of the appropriate aggregation procedure can be expanded to include *what emissions?* How should individual pollutants in any class, such as hydrocarbons, be aggregated? And *which emitters?* Is the EPA concerned with the emissions from any particular car or with the average level of emissions from any particular group of cars?

These questions about emissions aggregation must be answered in order to produce any agreement, even in principle, on the emissions standards implied by the Clean Air Act and to give quantitative meaning to the term "on-the-road emissions." It can be assumed that the intention of the Clean Air Act is to provide standards that apply to on-the-road vehicle emissions rather than to test measurement emissions, and that the EPA considers the CVS Cycle the best procedure for simulating actual driving.

Looking at the problem in greater detail, a definition of on-the-road emissions requires the specification of seven different parameters:

1. *Pollutant species.* Is 90 percent reduction required of each of the several hundred hydrocarbons or just of the aggregate?

2. *Units of quantification.* Should the units encompass particular species (ethane) or groups of species (hydrocarbons) per mile or per second, or fractional concentrations of species in parts per million?

3. *Time.* Should average emissions rates be considered on a per minute, per day, or per year basis?

4. *Group of vehicles.* Should emissions standards apply to each individual vehicle, to the average of each appropriately defined vehicle model, to the average of the output of each manufacturer, or to the average of all vehicles?

5. *Geographical region.* Should emissions be averaged for vehicles in a particular part of the country (such as the Los Angeles basin) or for the whole nation?

6. *Driving patterns.* What is the actual average mix of slow and fast driving, frequency of stops, and other driving parameters that affect emissions, including vehicle maintenance and repair?

7. *External conditions.* What are the temperature, climate, topography, road conditions, and time of day under which the driving takes place?

In general, the more narrowly defined the categories are, the more stringent is the standard implied by a 90 percent reduction. For example, a 90 percent reduction in each of the hydrocarbon species is much more difficult to attain than a 90 percent reduction in the total mass of hydrocarbons. Similarly, a 90 percent reduction in instantaneous emissions or in emissions under each of the driving conditions (start-up, idle, cruise speed, etc.) is more difficult to achieve than a 90 percent reduction in *any* average over different driving conditions.

The specification of each of these parameters in the conceptual definition of on-the-road emissions is the EPA's responsibility. For the purpose of this analysis we use specifications that are conceptually simple and appear to be consistent with the intent of the Clean Air Act. We define the seven parameters as follows:

1. Each chemically distinguishable species
2. Mass per vehicle mile
3. One year (in any particular year of the vehicle's usable life, e.g., in the second year of a 1970 vehicle)
4. Average emissions of vehicles in each engine class (as defined in the *Federal Register*)
5. Average emissions throughout the country
6. The driving patterns averaged over all parts of the country
7. The conditions averaged over all parts of the country over one year

This is consistent with a relatively simple conceptual definition of the average *on-the-road emissions* for cars in a particular class: *the total mass of the chemical species that is emitted from all such vehicles in the country in one year divided by the total number of miles driven by those vehicles.*

With this definition, the quantity of on-the-road emissions is now a well-specified number (in grams per mile), even though the number may not be known at present with any degree of accuracy. Since the quantity of on-the-road emissions for 1970 was a well-specified number (for each chemical species and engine class), a 90 percent reduction in this number is equally well specified. This number (for each relevant chemical species) can be considered the "Clean Air Act intended emissions standard." In order for an individual car to meet this standard, it must be the case that if all other cars with the same nominal production and driving history were driven in the average fashion specified above, then the average on-the-road emissions would not exceed the intended standard of the Clean Air Act.

This definition of on-the-road emissions provides the basis for a de facto emissions standard. Any individual vehicle (or class of vehicles) will produce a well-defined level of on-the-road emissions (i.e., those average emissions produced when driven under the average conditions specified above). This car (or class of cars) will either pass or fail a particular emissions test. The de facto emissions standard associated with this test is the actual on-the-road emissions level that will be produced by a vehicle or class of vehicles that barely passes the test. It is clear that, depending on the way the test and enforcement procedures are put together, the de facto standard can differ substantially from the nominal standard—e.g., 0.41 grams per mile (gm/mi) HC—to which the test score is compared.

Using this terminology, the rule-making task of the EPA is to define measurement test procedures and nominal emissions standards so that the de facto emissions standards are made equal to the Clean Air Act intended emissions standards. The enforcement procedure—which is designed to insure that the vehicle emissions do not exceed the nominal standards—would then insure that vehicle emissions do not exceed the Clean Air Act intended emissions standards. Of course, this correspondence is seriously complicated by the need to ensure that vehicle maintenance is performed at the level upon which the on-the-road standards are based. Special attention is given to this critical problem in the last section of this chapter and in Chapter 6.

Testing Scoring: Definitional and Empirical Problems of Measurement

The demands of actual measurement can affect the definition of emissions used in the test situation. Basically, there are two parts to the test scoring problem: definition of an unambiguous *measure* of emissions, and specification of an empirically unambiguous *measurement* of the emissions.

Before turning to the choice of measures and the specific problems of measurement presently faced by the EPA, it is worth considering what properties are desirable for emissions measures and measurements in general. In order to provide complete information, emissions measurements must be *instantaneous* in order to produce continuous records of the emissions. Such continuous records offer the best solution to the problem of test mode comparisons: emissions at each stage of a drive cycle would be independently known so that tests could be compared mathematically rather than empirically. Two properties of emissions measures are therefore desirable. First, they should be usable for *instantaneous* measurement, and second, they should be defined *independently* of any test procedure or instrument. In fact, the second property requires the first, since any noninstantaneous measure implies some procedure for averaging over time.

Independent Measures of Emissions. The first question that arises is whether a measure of pollutant emissions can be unambiguously defined independently of any specific test procedure or measuring instrument. The answer is yes; compared to other pollution variables such as toxicity or irritancy, gaseous pollutant emissions are ideal for defining an independent measure. Exhaust emissions are composed of discrete particles (molecules) that for any particular type of pollutant are identical. This allows the pollutant measure to be expressed in the most primary units possible: integral numbers and time. From a definitional viewpoint, emissions of CO are as easily countable as marbles rolling out of a vehicle tailpipe would be.[5]

What about the measures used by the EPA, such as grams of hydrocarbons per mile? Species can be aggregated in measures of emissions without introducing any ambiguity (although information is lost) as long as the class of species (such as hydrocarbons) is unambiguously defined. The expression of emissions in grams per vehicle mile is, for moving vehicles, just as well specified as numbers of molecules per unit time. Molecules of a given species have a well-defined average mass, and the number of molecules (or mass of molecules) escaping per unit of vehicle travel is also a well-defined quantity. Thus, a test-independent measure of grams per mile is a perfectly unambiguous measure of instantaneous emissions rates for moving sources.

As an instantaneous measure, however, units of grams per mile (unlike units of parts per million, used in earlier test procedures) cannot describe emissions from stationary vehicles (for which the instantaneous emissions rate per mile is infinite). In order to retain an instantaneous and test-independent emissions measure, the EPA would have to define an auxiliary measure (such as grams per second) that is applicable to stationary vehicles. The EPA has chosen to sacrifice the advantages of an independent measure, however, by using one that is test-dependent.

Test-Dependent Measures of Emissions. A test-independent measure of emissions is an instantaneous and unique measure for emissions. A CO emissions rate of 3.4 gm/mi has a unique meaning at any point in time. It describes all the relevant information about the emissions. An emissions standard could be written in terms of such a measure: e.g., CO emissions are never to exceed 3.4 gm/mi while the vehicle is moving. But once any averaging is introduced (by design or default) to the concept of an emissions measure, the measure can no longer be test-independent. The present CVS measure for emissions is such an averaged measurement.

---

[5]Unlike rolling marbles, molecules have the capacity to diffuse backwards and to react with other species. Back diffusion in a real tailpipe can be shown to be negligible, however, so that motion can in fact be considered unidirectional in the direction of the exhaust stream. The possibility of reaction requires that the location (distance down the exhaust stream) and the flow measurement be specified.

The measure of emissions used by the EPA in the CVS Test is not instantaneous, but involves an averaging procedure that is intrinsically defined in the test procedure. The measure is not really "grams per mile" but rather "total grams emitted during the test divided by total distance of the test." The measure is not unique as an independent measure of emissions; an infinite variety of emissions combinations (with emissions defined according to an independent measure) can produce the same CVS measure of emissions. When numbers are quoted from the CVS Test as measures of vehicle emissions, they are not meaningful without knowledge about the test procedure. In the present CVS measurement rules, the specification of the emissions measure and the specification of the conditions of measurement are combined in the definition of the emissions standard. This practice leads to the problem of test mode comparison.

Instrument-Dependent Measures of Emissions. The definition of a test-dependent measure of emissions does not rely on any specific method of measurement. For example, the CVS Test measure of CO emissions from a particular vehicle is independent of whether CO will be measured by infrared absorption or mass spectrometry. The question is, what meaning does an instrument-independent measure have if two instruments differ about its magnitude? This question is most conveniently answered by looking briefly at the larger problem of empirical measurement of emissions.

Pollutant molecules cannot be counted directly; their presence must be detected indirectly through some physical effect. The experimental measurement problem arises because of the gap between the actual pollutant, X, and the measurable effect, Y. The two problems are the correlation of X and Y and the absolute calibration of Y in terms of X. There are several reasons for the imperfect relationship between X and Y:

1. Interference: Y is affected by species other than X
2. Lack of sensitivity: Y is not produced strongly enough by X
3. Lack of precision: Y is not produced consistently by X
4. Long-time response: Y is produced after X.

The point to be emphasized here is that any pollutant measurement is indirect. The measurement sequence includes sampling, gas handling, physical observation, recording of observation, and instrument calibration. For indirect detection techniques, the additional steps of reactant preparation and reaction are required. Choice of measurement technique usually requires trade-offs among the sources of imperfection listed above.

Because the empirical measurement process cannot detect emissions directly by means of instrument-independent measures, instrument-dependent measures are often required. Instrument-dependent measures of the emissions X define the emissions in terms of an instrumental effect Y. This is the case for hydrocarbons in the CVS Test. Although the nominal CVS standards are defined in terms of the independent measure of mass per mile, the actual measure of hydrocarbons is instrument-dependent, based on the response of the flame ionization detector. The definitional measure of HC emissions for any particular hydrocarbon is no longer its mass per mile. The measure of the emissions is now defined as the contribution to the flame ionization current produced by that hydrocarbon multiplied by 0.945 times the equivalent propane mass per unit current response.[6]

There are two important drawbacks to instrument-dependent measures. First, because different instruments (even though they utilize the same techniques) produce different responses, a measure is ambiguous unless it is defined with respect to a single instrument. And the detection response of even a single instrument will change over time. This is a particularly troublesome factor for flame ionization detection (FID), since the instrument's relative response to individual hydrocarbons and its calibrating gas differ considerably with experimental conditions [Bruderreck, Schneider, and Halasz, 1964]. The second problem is that even if the instrument-dependent measure is unambiguous with respect to measurements made by a particular instrument, its relation to more universally understood quantities is not known. Unless all other quantitative discussion (for example, of health and smog effects) of that particular hydrocarbon is conducted in terms of that measure, there will be no way to compare the quantities. The problem is compounded by the presence of several species in the HC measure.

Given these complexities, the EPA's ultimate objectives for emissions measures should be twofold:

1. To develop instrumental techniques that can empirically resolve any measurement ambiguities that arise from the use of different test station instruments. This would enable the EPA to convert the present instrument-dependent measures into the independent measures in which the standards are expressed.

2. To develop instruments that can provide reproducible measurements of all emissions of interest in the test situation. These instruments could express their results in terms of independent measures by being calibrated against the unambiguous instrumental technique noted above. The ultimate goal for such test instruments would be to make possible instantaneous measurement of every chemically distinct species in the exhaust emissions.

Setting Standards for Enforcement

Given that emissions testing under the Clean Air Act must be done with less-than-perfect instruments and under test cycles that cannot possibly reproduce actual on-the-road conditions, there remains a great deal of latitude for the administrator of the EPA to define appropriate driving cycles, testing procedures, and test criteria. Effectively, by such rule-making, the administra-

---

[6]The EPA assumes that each carbon atom produces an equal contribution to the ion current and that the carbon to hydrogen ratio is 1:1.85. For propane this ratio is 1:2.67, which leads to the factor of 0.945.

tor has the authority to change the de facto standards imposed on manufacturers and car owners. Since the absolute value of the Clean Air Act intended standards is not known by the EPA administrator nor by anyone else, a wide range of testing rules and criteria can be considered consistent with some reasonable interpretation of the Clean Air Act. Someone must choose the measurement rules, thereby determining the de facto standards, and the EPA administrator is the person empowered by statute to make the choice. Rules and rule changes that fall within the bounds of present scientific uncertainty can therefore be thought of as discretionary policy instruments for the EPA.

Some of these instruments are less easily changed than others—particularly those requiring explicit alteration of rules already promulgated in the *Federal Register 36* (April 30) [1971]. Rules published in the *Federal Register* are not inviolate, however. Changes were made in the hot-cold weighting procedure and in HC and CO correction factors between November 10, 1970, when the original version of the 1973-74 standards was promulgated, and July 2, 1971, when the revised procedures were promulgated. Admittedly, such changes would not be so easy to make as those requiring mere reinterpretation of the rules already in existence. Correction-factor rule changes are generally irreversible, since within the limited context of comparing test emissions to test scores such changes *are* scientifically testable.

There are three components in the rule-making for enforcement of the Clean Air Act. The first is the establishment of the test measurement process itself and involves all the problems of measures, instrumentation, and driving schedules discussed earlier. The second part of the rule-making is the determination of criteria for vehicle sampling and aggregation, including sample size and possible methods of averaging. And the third component is the setting of nominal numerical test standards, which do not generally correspond to the desired target levels of actual on-the-road emissions and which are therefore subject to discretion. Of course, each of these three applies to prototype testing for certification, to assembly-line testing, and to the guidelines to be established by the EPA administrator for authorization of state inspections under the warranty provisions of the Clean Air Act.

Test Procedures and Measurement Rules. A group of seven test procedures and measurement rules needs to be considered: dynamometer driving schedule, ambient temperature, sampling procedure, classification of pollutant species, measures of emissions, correction factors, and mileage accumulation procedure.

In the *dynamometer driving schedule* vehicles are driven on a dynamometer through a driving schedule (speed vs. time) that is intended to reflect on-the-road driving conditions. The driving schedule used in the CVS Test was derived from studies of average driving patterns. [System Development Corporation 1971]. It is 7.5 miles in length and takes 1,369 seconds to complete. The final test score is affected by several details, which are specified in the *Federal*

Register 36 (April 30) [1971]. Because catalytic mufflers for removing HC and CO operate much less efficiently at low temperatures, the driving schedule must also specify the previous driving conditions. In particular, if the 7.5-mile average driving pattern began within a few hours of previous vehicle use, the engine and catalysts would still be at higher than ambient temperature, so that the emissions during warm-up would be much lower than those for a cold start (in which all parts of the vehicle are at the ambient temperature). A hot start-cold start weighting procedure was introduced to account for this difference. In the July 2 rules, the driving schedule was changed to include two parts: one complete, cold start drive through the 1,369-second schedule, and, after ten minutes, a second drive through the final 505 seconds of the schedule. The results of the two drives are averaged according to the weighting procedure.

Accurate calculations of what difference the hot start-cold start rule change made in the final test score for an average vehicle are not available outside the EPA. A rough estimate can be made, however, by assuming that data in the sample calculation supplied in the *Federal Register 36* (April 30) [1971] are representative of actual test results. The data supplied for HC are:

The total mass for "transient" (first 505 sec) portion of cold start schedule
= 4.03 gm
The total mass for "stabilized" (last 864 sec) portion of cold start schedule
= 0.62 gm
The total mass for "transient" (first 505 sec) portion of hot start schedule
= 0.51 gm.

Before the hot start-cold start rule change, the emissions test score would have been calculated as (4.03 + 0.62)/7.5 = 0.62 gm/mi. After the rule change, the test score would be recorded as (0.43 x 4.03 + 0.57 x 0.51 + 0.62) /7.5 = 0.35 gm/mi. This difference is considerable. The pre-rule change test score would have been 100(0.62 - 0.35)/0.35 = 77 percent higher than the post-rule change test score. The change in de facto standards was not quite this great because the nominal standards were adjusted to compensate partially for this change. The extent of numerical change in the 1975-76 standards from those which were originally considered to be a 90 percent reduction in the 1970 emissions is not known, but can be inferred, however, from the compensatory numerical change in the 1973-74 standards proposed by the EPA in the *Federal Register 36* (April 30) [1971]. The proposed change in 1973-74 nominal standards for HC was from 3.4 gm/mi to 3.0 gm/mi. The change with respect to the new nominal standards is thus 100(3.4 - 3.0)/3.0 = 13.3 percent. These calculations suggest that if the *Federal Register* data are representative, then the cold start-hot start rule change produced a relaxation of the de facto standards by about 64 percent, calculated with respect to the new de facto standards. The effects of this rule change on the three pollutant categories were calculated by the same method, and the results are listed in Table 5-1. These calculations could easily be repeated by the EPA, using more accurate data.

Table 5-1. Estimated Net Change in De Facto Standards Produced by Hot Start-Cold Start Rule Change

Species	Change in Test Score (Percent)	Compensating Change Made in Nominal Standards (Percent)	Net Change in De Facto Standards (Percent)
HC	-77	+13	-64
CO	-57	+39	-18
NO_x	0	+ 3	+ 3

Note: Data taken from *Federal Register 36* (April 30) [1971]. (Percentages calculated with respect to the new level.)

The change in the hot start-cold start rule has already been made. Calculations of its effects are included here to illustrate the magnitude of the changes in de facto standards that are produced by such rule changes. Presumably, the driving schedule weighting procedure could still be changed, with similar effects on the magnitude of the de facto standards. Or changes might be made in the driving schedule itself, such as shortening the driving schedule, increasing the weighting of the initial periods after cold start, or increasing idling time, all of which would increase the stringency of the de facto standards.

The *ambient temperature* of the test facility affects (1) the temperature of the initial cold start, (2) the time taken for the vehicle engine and catalysts to reach their stable phases, and (3) the actual temperature of the stable phase. The ambient temperature during the tests is presently required to be between 68° and 86°F. Although the magnitude of the test score change associated with a change in temperature is not publicly known, it is expected that a reduction in ambient temperature would produce an increase in the test score and therefore an increase in the stringency of the de facto standards.

The present *sampling procedure* employs constant volume sampling. From a scientific viewpoint, by making the test score reflect the actual emissions during the driving schedule, the CVS procedure is a significant improvement over the pre-1970 procedure. The previous procedure measured the fractional concentration of pollutants in the exhaust gas and then converted the results to mass per mile by using an assumed average flow of exhaust gases. But, the CVS method still does not produce a true measure of drive-cycle emissions. The measurement is not made instantaneously, as the gas emerges from the vehicle, but after the exhaust gas has been diluted and retained in a sample bag for some time. As noted previously, this sample bag approach introduces an averaging method, making the measure test-dependent. In addition, the measure does not record the actual emissions because the pollutant gases react during the sampling period. The gases from the initial stages of the driving schedule can be in the bag up to eighteen minutes before being measured.[7] During this time almost all the possible reactions

---

[7] That is, 505 seconds during the driving schedule and up to ten minutes after the test is completed.

(such as wall reactions, oxidation of CO and HC, reaction of NO_x with HC) reduce the final amount of measured pollutants. Because of this phenomenon, almost all changes in sampling procedure that move toward instantaneous emissions measurement serve to increase the stringency of the de facto standards.

There is considerable scope for reinterpretation and *reclassification of the types of pollutant species*. The question is what species the writers of the Clean Air Act had in mind when they specified a 90 percent reduction. Because different species differ in their pollution effects per unit mass, it is possible, for example, that a 90 percent reduction in total HC mass for some vehicles could be achieved while actually increasing the emitted mass of the most damaging HC species. If the 90 percent reduction were reinterpreted as referring to each individual HC species, the stringency of the de facto standards would be greatly increased.

Another problem is whether partially oxidized hydrocarbons (aldehydes) should be included in the HC standards. Gram for gram, aldehydes are considerably more polluting than most pure hydrocarbons (as discussed in Chapter 7). In fact, before most hydrocarbons become real atmospheric pollutants they must be oxidized. One automobile manufacturer's chemist has estimated that the mass of aldehydes (primarily formaldehyde) could be over 10 percent of that of hydrocarbons (and therefore have pollution effects comparable to the total mass of unoxidized hydrocarbons).[8] Thus, inclusion of partially oxidized hydrocarbons (aldehydes) in the HC classification would increase the stringency of the de facto standard.

It is fairly certain (as pointed out in Chapter 7) that the two lightest saturated hydrocarbons, methane and ethane, have no significant pollution effects (or, at least, effects that are insignificant compared with the effects of the heavier—especially unsaturated—hydrocarbons). It is these unreactive hydrocarbons that are the most difficult to remove by catalysts. For some catalysts the residual mass of HC emissions can contain up to one-half methane and ethane. Thus, the exemption of unreactive methane and ethane would considerably decrease the stringency of the de facto HC standards.

The question of variable weighting of species can also be applied to the NO_x category. At one time it was believed that most of the NO in exhaust gases was converted to NO_2 by reaction with oxygen. Since NO itself is thought to be less harmful than NO_2 (and may even reduce the reactivity of the HC emissions [Altshuller and Bufalini 1971] and remove harmful ozone), and since NO in low concentrations does not react rapidly with oxygen to form NO_2, it has been suggested that NO be excluded from the NO_x category. A less drastic change would be to calculate the actual NO_x mass (including both NO and NO_2) rather than to treat the NO component as if it were in the form of NO_2 (which is 53 percent heavier per molecule than NO). Thus, changes in the

---

[8]Since this estimate is still highly speculative, the industry employee who supplied it wishes to remain anonymous.

NO_x mass calculation rules that count NO at its true mass or that exempt NO altogether would decrease the stringency of the de facto NO_x standards. This would even be true if the nominal standards were adjusted to make them consistent with the newly defined NO_x, since the $NO:NO_2$ ratio in auto exhausts probably increases as the NO_x emissions level is reduced.

Based on this analysis, we consider the following recommendations to be consistent with the intent of the Clean Air Act. Standards should be applied to each distinguishable chemical species deemed to be a pollutant under the Clean Air Act. Aldehydes should be included, but methane and ethane should be excluded, and nitric oxide should be measured at its true mass. The probable net effect of these changes would be simultaneously to reduce harmful pollution and to relax the stringency of the overall standards and therefore reduce the costs of pollution control while increasing the benefits.

Earlier it was shown that the CVS Test uses an instrument-dependent measure of HC emissions. If this were changed to a "true mass" (independent) *measure of emissions,* the HC test scores would change. Whether the scores would go up or down depends on the relation of the true mass response of the present FID instruments to HC size and on the accuracy of the currently assumed C:H ratio of 1:1.85. Examination of FID response curves [Bruderreck, Schneider, and Halasz 1964] suggests that the net change produced in de facto emissions standards by converting to a true mass measure for hydrocarbons might be as high as 20 percent, although the direction of change is not yet known.

Two *correction factors* were explicitly included in the July 2 *Federal Register* rules that were not in the November 10 rules. These correction factors accounted for the background concentrations of detectable species in the diluent gas and for the interference of H_2O and CO_2 in the CO absorption (nondispersive infrared) detection. The net effect of these two correction factors, as well as the humidity correction factor for NO_x, was to make the de facto standards for HC, CO, and NO_x more lenient by 10, 5, and 6 percent, respectively. Such corrections factors are scientifically determined (since they relate the actual test emissions to the test score) and are relatively small. Other, as yet undetermined, correction factors for the calculation of details such as the dilution factor and other detection interference are likely to be less than 10 percent in either direction.

The 1975-76 standards require that vehicles meet the emissions standards after 50,000 miles use. The present procedure requires that an emissions test be carried out on representative prototypes that have been driven up to 50,000 miles and and their best records compared with those for the same prototypes at 4,000 miles to calculate the deterioration factor. The final test score for each specific model prototype will be determined by multiplying the zero-mile test record by the deterioration factor. Thus, any change in the mileage accumulation procedure that affects the difference between 4,000-mile and 50,000-mile test records will directly affect the initial and final test records for all the individual prototypes.

The most important determinants of the 50,000-mile test score will be the amount and timing of vehicle tuning and the driving conditions (speed, acceleration, road conditions, temperature, etc.) under which the miles are accumulated. The present rules specify the amount of tuning allowed (a fairly imprecise concept) but give the manufacturers almost complete freedom in choosing the driving conditions for mileage accumulation. Changes in the mileage accumulation rules that permit more tuning and parts (especially catalyst)[9] replacement, therefore, will make the de facto standards considerably more stringent. The adequacy of the current mileage accumulation schedule in simulating in-use conditions is considered in detail in Chapter 6.

The magnitudes of all these effects depend on the specific design of engines and catalysts. More exact quantitative calculations could be made by EPA officials when the data become available. Of course, the effective change in de facto standards depends on whether or not, and to what extent, nominal standards and test criteria are changed. More will be said about this in the next two sections.

Vehicle Sampling and Averaging. Under any given test procedure (at certification, at the assembly line, or on the road), vehicles are tested, and the results are evaluated by a pass-fail criterion. This criterion may take many forms. For example, every car may be required to meet some numerical standard; the average of some group of cars may be required to meet some standard; or, say, 90 or 95 percent of all vehicles may be required to meet a standard.

The stringency of the standard depends on the procedure applied. We will show in the next section that a given numerical standard applied to each car individually is far more stringent than the same numerical standard applied to the average of a group of cars. Furthermore, the variability of a single test measurement is far greater than the variability of a measured average, so that the probabilities of errors of commission (failing a car that should have passed) and of errors of omission (passing a car that should have failed) increase if standards are applied on a car-by-car basis.

Like the methods of aggregating measurements of individual vehicles, the rules for aggregating several measurements on the same vehicle can affect the stringency of the effective standards. In particular, if a car is allowed to be retested after an unsatisfactory initial test, the rules used to aggregate the two measurements have a substantial effect on the stringency of the standards. This issue will be considered later in a more detailed discussion of sampling and averaging. All of these issues of vehicle sampling and averaging will also reappear in our discussions of prototype testing, assembly-line testing, and on-the-road testing.

Numerical Nominal Standards. After the test procedures, sampling methods, and measurement aggregation criteria have been established for a given

---

[9] The EPA ruled [*Federal Register 37* 1972] on November 8, 1972, that a catalyst may be replaced at 25,000 miles.

test, a numerical nominal standard must be established. The vehicle scores (whether individual or aggregate) are to be compared with some nominal value in order to determine whether a vehicle or group of vehicles has passed or failed.

The key point is that the nominal standard applied should depend not only on the test procedures, but also on the sampling procedure and vehicle aggregation procedure. In order to achieve the same de facto standards, it will be necessary to use different nominal test standards depending on whether the standard applies to the scores of individual vehicles, averages, minimum or maximum scores on repeated tests, or to some other sample statistics. The next section deals with these issues; it is sufficient here to say that the choice of a nominal standard for a particular test cannot be made without knowing the characteristics of the sampling procedures as well as the characteristics of the physical test itself.

The Clean Air Act calls for a 90 percent reduction from 1970 emissions levels. This reduction can only be reasonably interpreted to apply to actual emissions on the road, or effective standards, not to numerical nominal standards. The latter logically depend on a number of parameters that the Clean Air Act does not explicitly define and that are left to the discretion of the administrator of the EPA.

One final point about vehicle deterioration is that a nominal standard applied, say, at the assembly line will affect actual on-the-road emissions in a way that will depend on deterioration and maintenance. Assumptions about deterioration and maintenance will therefore be critical in setting nominal assembly-line standards so that on-the-road emissions at 50,000 miles are at the desired levels. In short, the nominal standard applied at zero miles under a given test would be numerically lower than the nominal standard applied at 50,000 miles under the same test. How much lower it would be depends on assumptions about deterioration and maintenance.

CONCEPTS USED IN THE ANALYSIS

Manufacturers' Response to Emissions Testing

As stressed at the beginning of this chapter, the auto manufacturers are key decision-making units in air quality control policy, and their response to federal emissions testing and enforcement is a major determinant of the success or failure of control policy. In order to develop the concepts used in the analysis of the industry response, we make two assumptions. The first is that the threat of enforcement by the EPA is credible to the automobile manufacturers. If this is not the case, then no testing program can possibly accomplish its desired goals. The second assumption is that manufacturers can reduce the emissions of their products to meet whatever standards are in effect, provided that they spend sufficient amounts of money. Whether or not this assumption is valid for the 1975-76 standards as they now stand is not important; if it is found that manufacturers cannot possibly meet these standards, no matter how much they spend, then

84 *Clearing the Air*

surely those standards will be changed, or else the threat of enforcement will lose its credibility.

Under these assumptions, manufacturers can be expected to produce vehicles with average emissions below the standards. The precision of the testing procedures, their correlations with each other, the extent to which averaging is employed, and the ability of manufacturers to reduce variability in mass production will determine how much below the standards actual emissions will be.

The key factor in this regard is the variation between the measured test result and the actual emissions of vehicles being produced. This variation comes from several sources: normal variation in mass production, "green engine" variation (assembly line only), imprecision of test measurements, imperfect correlation between tests, and uncertainty about durability. Because of all of these factors, manufacturers face uncertainty about what the test scores will be, and they will therefore tend to design their vehicles to hedge against this uncertainty.

To see the argument that leads to this conclusion, consider the graph in Figure 5-2. The CO emissions for a particular line of vehicles have been

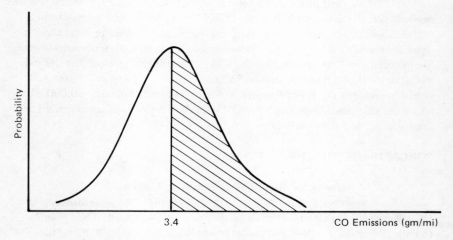

Figure 5-2. Variation in Observed Test Result

designed, say, to average 3.4 gm/mi, the 1975 standard. The test result, however, is not certain to be exactly 3.4 gm/mi. First of all, the vehicles tested will differ as a result of production variability. Furthermore, the test results will be uncertain because of lack of precision and poor correlation with other tests.[10] Thus,

---

[10]For discussion of this point, see the section in this chapter on "The Meaning and Significance of Correlation."

the observed test result will vary according to a probability distribution something like the one shown in Figure 5-2. The degree of spread will depend on the degree of mass production variability, test precision, correlation, and the extent to which averaging of vehicles is used in computing the test score.[11]

If manufacturers face the uncertainty shown in Figure 5-2, they surely will not design their vehicles to emit, on average, 3.4 gm/mi CO. If they did, the test score would come out above 3.4 gm/mi with a probability of 1/2, represented by the shaded area in Figure 5-2 (where the total area under the curve represents a probability of one). Therefore, they would have a 50 percent chance of failing the test, and that would be too much of a risk for them to take.

The response of manufacturers to such uncertainty is clear, as corroborated by widely applied methods of quality control. They will seek, simultaneously, to shift the graph of Figure 5-2 to the left by reducing average emissions below the standard, and to reduce the spread by decreasing production variability. Thus, the probability distribution of test results will shift, as shown in Figure 5-3. With the vehicles designed to emit only 2.8 gm/mi,[12] on average, and with the variability reduced, the probability of failing the test (shown by the shaded area in Figure 5-3) will be reduced to a level acceptable to the manufacturers.

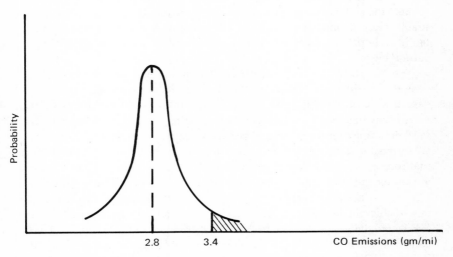

Figure 5-3. Likely Manufacturer Response to Test Variation

---

[11]The spread is also strongly affected by the retesting option available under the current federal procedure. This complication is introduced later.

[12]The numbers used here are hypothetical; some actual numerical results are given in the following sections.

In particular, a rational manufacturer will select his target emissions levels so that the marginal cost (in terms of expenditure on vehicle design and production) of reducing the probability of failing under any given test is equal to the cost (in terms of loss of face and lost profits) of failing the test.[13] In most cases, since the cost of failure is high, this will lead to a very low acceptable probability of failure and, therefore, an effective standard that is actually well below the nominal standard.

Nominal Standards, Averaging, and Sampling

If one were to do a complete analysis of the optimal air pollution control policy, including automotive testing, then one would naturally have to consider the social costs of air pollution (health effects, etc.) and weigh them against the costs of control that result from the policy. The result of such a comprehensive analysis would be an optimal set of air pollution levels, which could then be translated into optimal automobile emissions levels. It is the responsibility of Congress to make the value judgments necessary to determine the optimal level of air quality. The standards set in the Clean Air Act presumably represent the target levels of ambient air quality deemed by Congress to be in the nation's interest. Consequently, the corresponding reductions in automobile emissions (i.e., the 90 percent reductions for 1975-76), which would be necessary to meet the ambient air quality targets, are Congress's estimates of the most desirable levels of emissions reduction. The automobile emissions standards established in the legislation are therefore the target levels of emissions from automobiles.

The arguments presented earlier in this section (and the Appendix) demonstrate, however, that if the threat of enforcement is indeed credible, if the penalties are sufficiently great, and if it is possible to meet the standards, then—if there is any uncertainty whatsoever in the testing procedure—manufacturers will set their average emissions targets below the legislated levels. But if those levels are optimal in cost-benefit terms, then it would be inefficient to establish a policy that forces actual emissions below those levels. Such a policy would force expenditures on emissions control over what Congress determined to be justified.

There are two ways to alleviate this situation. The first is actually to raise the nominal standard used for enforcement and testing. Consider Figure 5-4, where the nominal standard for CO has been raised to 3.8 gm/mi. A probability of failure acceptable to manufacturers could then be achieved, in this hypothetical example, by designing vehicles to emit at precisely the target level, 3.4 gm/mi.

In reality, it may be difficult to change the nominal standards at this time, by changing the test procedure or by any other means, but such a change —if intended to result in emissions at, rather than below, the target levels—would

---

[13]The argument is presented in more detail in the appendix to this chapter.

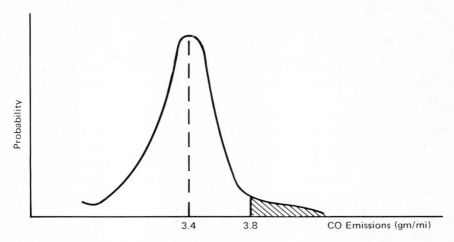

Figure 5-4. Adjustment of Nominal Standard To Achieve Target Level

clearly meet the intent of Congress. In summary, conscientious enforcement, through testing, of 1975-76 automobile emissions standards which represent 90 percent reductions from 1970 levels will result in actual emissions below the target levels. It may be desirable for nominal standards to be set at a somewhat higher level so that *actual emissions* will be near, and not below, the 90 percent reduction levels.

The second remedy available, and one that would alleviate the necessity for raising the nominal standard and for second-guessing the response of manufacturers to an uncertain testing situation, is to reduce the variability in testing by applying the standards to averages of vehicle emissions rather than to individual vehicles. This applies to prototype testing, assembly-line testing, and on-the-road testing for purposes of recall. The average of the emissions of 100 vehicles has 1/100 the variance of the emissions of a single vehicle. If averaging were used, the spread in Figure 5-3 would be much smaller, and the discrepancy between the nominal standard and the effective standard would be minimal.

Averaging is clearly supported by Congressional intent in the Clean Air Act. The intent of the Clean Air Act emissions standards is to improve ambient air quality by 90 percent in 1975; therefore, the standards must be interpreted as applying to *average* emissions of all vehicles.

These arguments for averaging have been given in the case of a single-pollutant world. Averaging in testing is even more desirable when, in actuality, there are three classes of pollutants. The amounts emitted will generally be correlated with each other, and the correlation could conceivably be either positive or negative. It will tend to be positive between HC and CO, and negative between

NO_x and each of the other two, because of the chemical and physical trade-offs inherent in the internal combustion engine. If, for example, it were cheaper to design half of all cars to emit .6 gm/mi HC and .2 gm/mi NO_x, and half to emit .2 gm/mi HC and .6 gm/mi NO_x, than to design all cars to emit .4 gm/mi HC and .4 gm/mi NO_x, then the former should be encouraged. Without averaging and with random selection of test vehicles, however, manufacturers would tend to choose the latter option.[14]

In summary, the intent of Congress in the Clean Air Act was to improve ambient air quality throughout the country. Since the overall ambient air pollution attributable to automobiles is a function of the *average* emissions of automobiles (for a given number of automobiles), the emissions standards to be enforced by testing should, where possible, be based on *averages* of vehicle emissions, rather than on *individual* vehicle emissions. Applying test standards to averages, in addition to meeting the intent of the legislation, has the advantage of reducing the effective variability of the test procedure and thereby minimizing the discrepancy between the nominal standards applied and the actual emissions produced in response to those standards.

The objective used throughout this chapter is the minimization of costs, subject to the constraint that actual emissions satisfy the standards on average. In political reality, another constraint might require that the nominal standards as they now exist not be changed, at least for prototype testing. Unfortunately, the first objective, as stated above, is not entirely credible. In a world of uncertainty, we can never be 100 percent sure that we are going to meet the standards. Clearly, Congress could not have intended that degree of confidence, which would necessitate expected ambient pollution levels essentially at zero; effectively, we would have to stop driving. A sensible way of interpreting the legislation, therefore, is that emissions should be expected to meet those standards.[15] Thus our objective is to minimize cost, subject to the constraint that *expected* emissions are at or below the standards set by Congress and possibly subject to the constraint that the nominal standards cannot be changed. The costs to be minimized include costs of emissions control to manufacturers (passed on or otherwise), costs of testing (borne by manufacturers or federal, state, or local governments), and costs of maintenance and repair to owners (or to manufacturers under the warranty).

Testing policy should be formulated, therefore, with these factors in mind. The selection of the number of vehicles tested—at the prototype stage, at the assembly line, and on the road—is a policy decision that can influence the overall variability of the testing procedure and therefore affect the costs of achieving emissions levels safely below the standards. Testing enough vehicles and aver-

---

[14]Of course, steps would have to be taken to insure that the two types of vehicles were not unequally distributed throughout the country.

[15]The analytical assumption of risk neutrality is probably justified at the low levels of emissions with which we are dealing.

aging the measurements would reduce testing variability and translate into savings to consumers. The size of the sample should be large enough so as not to force manufacturers to overspend on reducing emissions below the 1975 target levels because of uncertainty about passing the test, but small enough to keep actual testing costs in line with the benefits of reduced variability.

An additional complication arises if vehicle retesting is permitted, as is currently the case in prototype testing. If a vehicle is allowed to be retested (whether after repairs and adjustments are made or not), and if the best score thus generated is the score to be compared with a standard, then the implied effective standard is less stringent than would be the case if a vehicle could be tested only once. In fact, if the number of retests is not fixed in advance, then a procedure that uses the average of the test scores for the vehicle is less stringent than a comparable procedure using a fixed number of test trials. This is because a manufacturer (or car owner) might elect to "quit while he is ahead" or continue to test until the average is near or under the standard.

This is not to say that retesting should not be permitted. We do maintain, however, that the nominal standards for any procedure in which retesting is permitted should be numerically lower than for a comparable procedure without retesting, if the effective standards being enforced are the same. Furthermore, the standards for a procedure in which the retests are averaged should be numerically lower than for a procedure in which the minimum score is used, and the standards should be even lower if the number of retests is not specified in advance.

The analytical complications introduced by retesting are great. The setting of sensible nominal standards for procedures with complicated and ill-defined retesting options (such as the existing certification procedure) is difficult if not impossible. The only rationale for allowing retesting is to counteract the high measurement variability of a test which each vehicle is required to pass. By allowing retests, the effective emissions standards are made less stringent, counteracting the effect of measurement error (which makes the effective standards more stringent). By averaging, rather than by introducing scientifically unsound retesting options, the same result can be achieved without losing sight of the effective emissions standard being enforced. For tests that vehicles must pass individually (e.g., screening for high emitters at the assembly line and on-the-road inspection), retesting will be required, but the nominal standards should be lowered accordingly.

Thus, test procedures that use retesting options as a means of counteracting test variability should be accompanied by nominal standards calculated so that the desired effective standards are enforced. In view of the complexity and arbitrariness of this approach, retesting should be allowed only when vehicles are required to pass a test individually; in other cases, averaging without retesting should be used.

The Meaning and Significance of Correlation

There is general consensus that tests used at different points—from certification to assembly-line testing to field testing—ought to "correlate" well. That word is used in the legislation, and it has led to a good deal of controversy and confusion over whether or not *any* test correlates with the federal certification test.

As shown in the Appendix, the notion of statistical correlation, in the usual sense, is less useful than the notion of *residual variation* in a test after its score is adjusted to correspond to the CVS score. If we could perfectly predict the measurement of the emissions under some test, given the measurement from the federal CVS Test, then for all intents and purposes those tests would correlate perfectly: there would be no residual variation in the test considered once its measurement values had been adjusted to correspond to CVS Test values.

For example, if "true CVS" CO emissions were x gm/mi and a particular test registered a score of $y = 2x - 3$, then, given the test score y, we could compute with certainty the "true CVS" emissions $x = (y + 3)/2$. If, however, the test registered the score $y = 2x - 2$ with probability 1/2 and the score $y = 2x - 4$ with probability 1/2, then, given an observation y, the true emissions x could not be computed with certainty. Rather, we could only conclude that $x = (y + 2)/2$, with probability 1/2, or else $x = (y + 4)/2$, also with probability 1/2. In the first case, the tests correlate perfectly; in the second case, they do not.[16]

For the analysis that follows, especially in the sections dealing with assembly-line testing and on-the-road testing, it is convenient and valid to consider all test scores in their CVS-adjusted form. The total variation in a test consists, then, of measurement error and process error in converting from one test scale to the other. The latter is essentially the variation inherent in the testing process relative to the federal certification test: it is the error that would have been predicted in a vehicle's "true" score under the test (i.e., without measurement error) from its score under the CVS Test (without measurement error). The Appendix defines and develops these concepts in more detail.

Deterioration and Maintenance

The Clean Air Act target emissions levels are meant to apply throughout the useful life of a vehicle. In particular, "average lifetime emissions," or emissions at 50,000 miles, must meet the standards [*Federal Register, 35* (November 10) 1970]. These average lifetime emissions are not the same as the

---

[16]It should be noted that, because of measurement error, a test does not correlate perfectly with itself! In other words, we cannot predict with certainty the test score of a vehicle in one test trial from its score on another trial.

emissions of a new vehicle or a vehicle with 4,000 miles because of deterioration and aging of the vehicle and because of the need for maintenance.

It is conceptually useful to divide the emissions history of a vehicle into two parts. The first is the increase in emissions that occurs over time as a result of the aging of the vehicle, assuming that perfect and continuous maintenance is performed. Included in such maintenance is the replacement of minor engine parts (spark plugs, points, etc.), the adjustment of engine parameters (ignition timing, idle rpm), the adjustment of the emissions control system, and the replacement of emissions control parts such as catalysts and PCV valves. The replacement of the carburetor or of any major part that is not normally replaced during the useful life of a car is not included. A typical graph of this "perfect maintenance" profile is represented by the dashed line in Figure 5-5. The deterioration is typically gradual and is due to normal wear and tear and the general aging of the system.

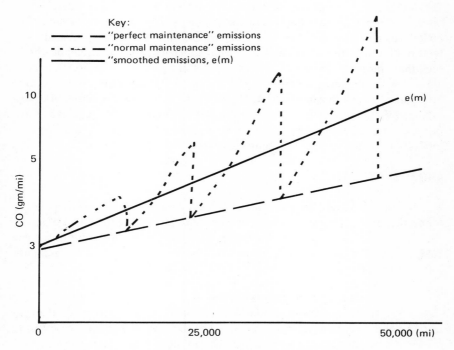

Figure 5-5. Typical Patterns of Emissions Deterioration

Of course, no vehicle undergoes continuous, perfect maintenance. In fact, for any reasonable maintenance schedule, emissions will rise steadily after maintenance until the next service is performed, reducing emissions toward the perfect maintenance level again. The dotted curve in Figure 5-5 illustrates a typical "normal maintenance" emissions profile for CO. Note that for NO_x, under certain technologies, emissions may actually fall below the perfect maintenance level because a poorly tuned engine may emit less NO_x than a well-tuned engine. As drawn, Figure 5-5 assumes that maintenance occurs every 12,500 miles.

The deterioration profile that is relevant for the application of the Clean Air Act "useful life" standards is obtained by smoothing out the dotted curve in Figure 5-5. This "smoothed" profile is represented by the solid line in Figure 5-5, and it is labeled $e(m)$. Note that if the between-maintenance deterioration in each of the four intervals had been of the same size, the smoothed profile would have been parallel to the perfect maintenance profile. In the figure, between-maintenance deterioration is shown to be progressively greater with age, so that the slope of the smoothed (or average) deterioration profile is greater than the slope of the perfect maintenance profile.

There are, therefore, two elements in the enforcement of the Clean Air Act useful life standards. First, manufacturers must produce cars that will, on the average, have emissions below the standards at 50,000 miles, assuming maintenance determined to be reasonable by the EPA. Thus, the combined effects of built-in deterioration and between-maintenance deterioration must be such that, under reasonable maintenance, the average emissions at 50,000 miles are under the standards. The second element of the enforcement problem is to insure that the maintenance is actually performed by car owners. Otherwise (for example, if a catalyst fails at 10,000 miles and is never replaced), emissions may increase by a factor of ten even though the built-in deterioration (that is, deterioration not correctable by maintenance) is rather small.

PROTOTYPE TESTING

The one well-developed component of the testing package is the prototype test for certification. The regulations in the *Federal Register* have the effect of law, though they have already been changed significantly (in the treatment of hot and cold starts and of catalyst replacement) and can be changed again if the administrator of the EPA so chooses.

Rationale for Prototype Testing

If emissions at zero miles were the only important criterion, then all testing could be done at the assembly line. But since we care about the performance of a vehicle throughout its useful life, it is necessary to have a way to insure that cars are being built to provide average emissions in compliance with the standards and that proper maintenance is being performed by the owner. On-the-road inspections can in some circumstances accomplish the latter, but, as we will argue in this chapter's final section, they cannot adequately do the former. The EPA therefore tests prototypes and uses durability data on a small sample of vehicles to extrapolate the emissions of the test vehicles to 50,000 miles.

Prototype testing is clearly necessary for the testing of durability, which obviously cannot be done at the assembly line. Prototype testing, however, should not be relied upon as the sole means of verifying that new vehicles are meeting emissions standards. First of all, new vehicles at the assembly line may bear little resemblance to prototype vehicles submitted by the manufacturers for certification. Second, the ad hoc retesting options currently in operation in prototype testing for certification[17] obscure the effective standards being enforced by the certification procedure, for reasons given in the previous section. Since prototype testing is necessary for the collection of durability data, such data could then be used at the assembly line to extrapolate new-vehicle emissions to 50,000 miles in the manner described in the section on "Assembly-Line Testing" and in the Appendix.

A Model of Prototype Testing

The model used in this analysis, and which applies to all three stages of testing with some modifications, is described here and sketched out in more detail in the Appendix. In what follows, consider a single engine type (model vehicle) and a single pollutant—say, CO. (The same general results hold in the three-pollutant case.) The points to be drawn from the analysis are the following:

1. Even with averaging, prototype testing is a high-variability procedure and is especially sensitive to the treatment of durability data.

2. As a result of the high variability, manufacturers can be forced to spend excessive amounts (as described in the previous section) on reducing emissions *below* target levels intended by the Congress and the EPA.

3. The procedures for collecting and processing durability data need careful reviewing, since certification and assembly-line test criteria may be extremely sensitive to these procedures.[18]

First, it can be shown that a criterion using averages is a viable alternative to that currently employed by the EPA for certification. The current procedure seems more stringent than an averaging procedure, since all of the vehicles tested must pass; in fact, it is more lenient because of the unlimited retesting opportunities available. An averaging procedure is less ad hoc and more in keeping with the objective of reducing overall emissions to the target levels as set by Congress and interpreted by the EPA.

A modified CVS procedure might be set up as follows: Select a random sample of vehicles for durability testing to get an estimate of the deterioration factor. Then measure a sample of test vehicles at 4,000 miles and extrapolate the observed emissions using the deterioration factor to estimate emissions at 50,000 miles. The variability of this testing procedure is calculated in the Appendix.

---

[17]At present, the EPA selects four vehicles for testing from among a pool of prototypes submitted by the manufacturer from each engine class. Each vehicle is required to pass the test. If a vehicle fails, however, the manufacturer may request a second test and submit a new vehicle to be tested. There is no definite point at which a model vehicle is considered to have irrevocably failed to obtain certification.

[18]This point is discussed in the next section.

Based on the following estimates of variability: production variability = ±20 percent, deterioration variability = ±50 percent, and CVS Test variability (at 1975 levels) = ±20 percent, the standard errors of estimates of CO emissions based on a sample of four durability vehicles and four test vehicles would, at 1975 levels, be about 0.75 gm/mi (compared to a standard of 3.4 gm/mi), or about 22 percent. Based on ten durability vehicles and ten test vehicles, the standard error drops to about 0.5 gm/mi, or about 15 percent. The more vehicles of both types included in the average, the smaller the error. Furthermore, these calculations are extremely sensitive to the number of durability vehicles used.

In designing and fabricating vehicles, manufacturers work (either explicitly or implicitly) with four performance parameters for each pollutant—the average and the variance of emissions, and the average and variance of deterioration in emissions. They will select these parameters so that the probability of failing the test is reduced to some small level, say 2 1/2 percent. As shown in the Appendix, based on the standard error just given, manufacturers seeking to reduce the chance of failing the test to about 2 1/2 percent will design vehicles to emit, on the average, only 2.4 gm/mi CO throughout their useful life, rather than the nominal standard of 3.4 gm/mi. If this does not happen (under a procedure without retesting options), then either the manufacturers are willing to take a higher risk of failure or the threat of enforcement is not credible to them.

Analysis behind the manufacturers' selection of their design parameters and their permissible probability of failure is also included in the Appendix. Under reasonable assumptions, in any case, manufacturers will be forced to design vehicles to emit *less* than the target standard because of uncertainty in measurement and production, provided that the *nominal* test standard is not somehow changed. If the nominal standard cannot be changed, then either an average with larger sample sizes should be employed or retesting should be allowed, but under a less ad hoc procedure in which the enforced standard is well understood. Careful statistical analysis would be required in order to implement the latter option effectively.

The choice of sample size at the prototype stage should be discussed before we turn to the critical issue of the use of durability vehicles. The objective in designing the test is to minimize costs, including the cost of testing and the cost to manufacturers and consumers of reducing emissions below the target levels. The constraint that emissions must be at least as low as the target levels has been shown to be not binding if the threat of enforcement is credible. Thus, as shown in the Appendix, the number of durability vehicles and emissions test vehicles should be chosen so that the cost of each type of test is just equal to the marginal reduction in production and design costs due to the reduced variability afforded by an additional test. Since the overall variability of the test procedure is so sensitive to the durability results, and since durability results may also, and should, be used to extrapolate assembly-line results to 50,000 miles, the use of more durability vehicles at the prototype stage should be considered.

Procedures and Processing for Durability Data

First of all, there is some confusion in the federal procedure, as it now stands, about the model of deterioration used to extrapolate test data to 50,000 miles. The statistical procedure used in estimating the deterioration rate assumes a *linear* process of deterioration, but this procedure is then used to generate a factor by which emissions at 4,000 miles are to be multiplied. If the linear model is used, rather than a logarithmic model (as used in the California surveillance study), then an additive constant, rather than a multiplicative factor, should be used for extrapolation. In any case, since the testing procedure is so sensitive to the treatment of durability data, the inconsistency should be corrected and the best possible model used.

More important still is the actual procedure by which durability data are collected. In particular, if the estimate of the deterioration factor is to reflect actual driving conditions and to predict actual lifetime vehicle emissions, then a more realistic mileage accumulation procedure should be used. Normal wear and tear, adverse weather conditions, and other factors should be reflected in the mileage accumulation procedure wherever possible. Otherwise, the estimate of the deterioration will underestimate lifetime emissions. Climatic and other factors that cannot be reproduced under test conditions should be incorporated by extrapolating the observed deterioration factor to average national climatic and driving conditions.[19]

Another aspect of the durability data procedure, which is critical in view of the possible use of catalytic mufflers on 1975 models, is the maintenance allowed during the test period. It is essential that the allowed maintenance in durability tests of prototypes be equivalent to the manufacturers' required maintenance under the warranty system. Otherwise, manufacturers could protect themselves from the warranty by requiring maintenance in excess of what is shown at certification to be necessary to meet, on average, the 50,000-mile standard. This is just one example of the strong interrelationship among the various stages of testing and the need for a comprehensive testing strategy.

ASSEMBLY-LINE TESTING

From the government's point of view, there are two motivations for required assembly-line testing. First, assembly-line testing could be used to force manufacturers to build new vehicles to conform, on the average, to emissions standards. If prototype testing were also used for enforcement, then assembly-line testing could complement prototype testing by making sure that the prototypes are indeed representative of the vehicles being produced for sale. Second, assembly-line testing could be used to protect car owners from purchasing an automobile

---

[19]Chapter 6 considers the consequences if it proves impossible to develop a true durability test of the type suggested here.

that will fail state inspection tests. This protection would be especially important if state inspections were in operation but there was no effective warranty to protect car owners. If there is a successfully implemented warranty, then manufacturers will have an incentive to screen out the highest emitters on their own.

Procedures and Criteria for Assembly-Line Testing

The main difficulty with assembly-line testing is the "green engine" problem—that is, the high variability in emissions from engines that have not yet been broken in. Thus, the use of averaging at the assembly line is absolutely essential, given the numerical (nominal) standards that now exist. In addition to requiring that the average emissions meet the standard, it may be desirable to protect car owners by screening for the highest emitters—by requiring, for example, that 99 percent of the cars not exceed some maximum emissions level well above the nominal standard.

There are essentially two ways to conduct assembly-line testing of average emissions. One way is to run a quality audit on a small sample of vehicles—say, 2 percent—using the CVS Test. In this case, there is no problem in correlating the assembly-line test and the prototype test. The other way is to use a somewhat larger sample and a less expensive test, such as an Idle Test, which does not correlate particularly well with the CVS Test. In this case, the sources of variation in measurement at the assembly line are fourfold: natural mass production variance, green engine variance, correlation error in translating results from the test used (e.g., Idle Test, Seven-Mode Test, Federal Short Test) into scores comparable to CVS scores, and measurement error.

The averages of zero-mile emissions at the assembly line clearly do not, however, provide enough information to determine whether emissions will meet the standards throughout the useful life of an average vehicle. The estimates of deterioration obtained from prototypes must be added to the observed assembly-line emissions (if a linear model is used) in order to extrapolate to 50,000 miles. Thus, for a vehicle to pass the assembly-line test, its average (CVS-adjusted) assembly-line emissions, plus deterioration, must be under the standards.

This procedure for extrapolating assembly-line measurements necessarily introduces much variability into the system, however, because the estimates of deterioration obtained from prototypes are extremely variable themselves. In addition, the procedure is very sensitive to the number of durability vehicles used to estimate the deterioration rate. The number of prototypes used for that purpose should therefore be chosen with care.

In spite of the great variability of the assembly-line testing procedures (green engines, poor test correlations, high-variance extrapolation), it is nonetheless the case that 100 percent testing at the assembly line, using any reasonable test, will most likely result in an insignificant degree of variability in the estimate of average emissions. This is, in fact, the power of averaging and can be demonstrated as follows: Suppose that an Idle Test has a correlation of $r = .1$

with the CVS Test. This is an entremely conservative estimate for HC and CO (values of $r = .5$ are to be expected) [Environmental Protection Agency 1972], but it may be approximately correct for NO_x, which is not measured accurately at idle. Nevertheless, we will consider measurements of CO in this example. Based on calculations given in the Appendix, the standard error of an average of 400 observations of CO emissions at zero miles, using a test that correlates as poorly as $r = .1$ with the CVS Test, would be only about 0.23 gm/mi, or 7 percent of the standard of 3.4 gm/mi. If the deterioration could be measured perfectly, then the standard error of the computed useful life emissions would still be 0.23 gm/mi, or 7 percent. If we base our extrapolation on, say, ten prototype durability vehicles, then the standard error doubles to about 0.45 gm/mi, or 13 percent.

It has been argued by manufacturers and others [Environmental Protection Agency 1972] that an assembly-line test can be said to correlate well with the CVS Test if, for a given batch of vehicles, the probability of failing the assembly-line test is no greater than the probability of failing the certification test. (Failure in this case means that average emissions exceed the numerical standard to which they are compared.) Further, however, the probability of failure (or, equivalently, the test confidence interval) should not only be at least as small as for the CVS Test but should in fact be not much smaller. This stipulation ensures a minimum testing cost and also makes the testing package more robust by preventing manufacturers from designing cars to pass the most binding test; cars would have to pass a battery of tests, resulting in a more honest picture of emissions.

A naive but justifiable approach might be to select the test and the sample size so that, for a fixed nominal "useful life" standard, the probability of failure at the assembly line is equal to the probability of failure at the prototype stage. As is shown in the Appendix, using the conservative estimates of correlation given above, well under 100 percent assembly-line testing would result in net variance as small as a CVS-prototype procedure using ten durability vehicles and ten emissions test vehicles.

In summary, therefore, if averaging is used, the variability of an Idle Test applied to all new vehicles will not be significant, even considering the green engine problem. An Idle Test applied to all vehicles at the assembly line would be a cheap and adequate means of verifying that vehicles being produced conform to emissions standards, without placing manufacturers in a position of having to overshoot the nominal standards by a great margin.

Quality Audit and Correction of Extreme Emitters

If the correlation between tests is so poor that 100 percent assembly-line Idle testing results in unacceptable overall test variance, then (assuming that the nominal standard cannot be raised) it may be preferable to use a 1 or 2 percent CVS quality audit at the assembly line, as California is now doing. In this case, the required sample size will be rather small, due to the low measurement error and the absence of a correlation problem if CVS is used. In fact, only the green engine problem will cause the variance to be higher than at the prototype stage. The choice between a 100 percent Idle Test and a 1 or 2 percent quality audit (assuming that such tests yield comparable confidence limits, on the average) depends on the relative costs. If the cost of a CVS Test is about $75 to $100 and if the cost of an Idle Test is about $1, this decision could be close. A sensible short-term policy might be to implement a CVS quality audit in addition to the Idle Test, since the data obtained could be used to learn more about inter-test correlation and to derive test score correction formulas.

One final point should be mentioned before we turn to on-the-road testing. If the warranty and recall are effective, manufacturers will have an incentive to correct the highest emitters at the assembly line, and this is both permissible and desirable, provided that retest scores do not enter the average. If the warranty and recall are not implemented or are not effective, then the EPA might require 100 percent assembly-line testing (even if a CVS quality audit were used for testing average emissions) and stipulate that no car (or possibly only 1 percent) might exceed some admissible maximum level. That maximum level should depend on the kinds of standards that owners are required to meet in state inspections, and on normal production variation.

ON-THE-ROAD TESTING, RECALL, AND WARRANTY

Objectives of On-the-Road Testing

Prototype and assembly-line testing cannot guarantee that actual emissions of in-use vehicles will be anywhere near the target levels. This is because deterioration as measured at certification may not be a valid representation of actual emissions except under a rigorous set of conditions. The mileage accumulation procedure used in durability testing must be a true simulation of road conditions, and the vehicles in use must receive the same quality of maintenance as frequently as in the prototype test procedure. For example, with a catalyst-controlled ICE it is likely that failure to perform service (including catalyst replacement) or actual tampering with control devices can easily yield emissions deterioration five to ten times that under proper service conditions. It seems hopeless to attempt to meet the Clean Air Act goals with a technology prone to deterioration without constant care, unless vehicle owners perform the service specified by the manufacturers and approved by the EPA. Since the motorists very likely will have ample incentives *not* to perform this service (to save money and to allow their cars to run better), strong inducements must be provided to insure that emissions control systems are cared for.

Thus, the primary objective of state inspections and associated programs of enforced maintenance is to provide this inducement. If the enforcement system is combined with a warranty (as provided in the Clean Air Act), owners will have an incentive to protect themselves against failing the inspection by performing adequate service. Naturally, the advisability of setting up inspection programs depends on the relative attractiveness of other means of holding down on-the-road emissions—such as a shift to an inherently stable technology, as discussed in Chapters 2, 3, and 4. It also depends on the capacity and willingness of the states to implement these schemes and to impose sufficient financial penalties to force motorists to take them seriously. In Chapter 6 we consider in more detail the problems of implementation as well as the consequences of a failure to make these systems work. Here, the focus is on the way such enforcement systems must function if they are to be effective components of a consistent testing strategy.

In addition to the primary objective of encouraging motorists to perform adequate service, recall and warranty together have a second objective of reinforcing the incentives on manufacturers to produce vehicles that are more durable under road conditions and that can be expected to meet the "useful life" standards, given normal deterioration and maintenance.[20] Other objectives, which in some cases set limits on the ability to achieve the first two, are not to penalize producers if the average emissions of a line of cars are, in fact, under the standards; not to penalize car owners if maintenance is performed which would, on the average, result in meeting the standards; and not to induce owners to perform maintenance in excess of what is necessary to meet the standards, on the average.

Recall

The recall provision of the Clean Air Act is designed to help achieve the second objective. It complements certification testing by requiring vehicles, up to five years or 50,000 miles, to meet the standards, on the average. The test used as a basis for recall could be the test used in state inspections or it could be a reapplication of the CVS Test on the road. By using a different test, for the reasons cited in the discussion of assembly-line testing, the overall testing package becomes more robust: manufacturers cannot design vehicles to pass a single test.

The same kind of analysis applied to assembly-line testing can be used here to show that, under reasonable assumptions about correlation of the recall test with the CVS Test, a sample of cars not much larger than the sample of prototypes, say 1 percent of the model line, will give confidence intervals at least as narrow as in the certification test. Note that, compared to the prototype test, the variance of the recall test is greater because of imperfect correlation and higher testing error; but some of the error encountered in prototype testing (because of the extrapolation of 4,000-mile data to 50,000 miles) is absent, since at recall the actual on-the-road emissions for the whole sample are measured.

For the recall provision to be meaningfully and equitably applied, only vehicles for which the manufacturer's specified maintenance has been per-

---

[20]It is argued in Chapters 3 and 6 that this link is weak, but clearly it should not be ignored.

formed should be considered. Limits on what constitutes reasonable maintenance will, of course, have to be established by the EPA. The recall procedure is technically feasible, then, provided that there is some way to verify that required maintenance has been properly performed.

If owners have no incentive to perform the required maintenance, either because of lax state standards or because of an ineffective or nonexistent warranty (which would be valid only if the required maintenance were performed), then the recall will be unenforceable. Thus the recall's effectiveness depends on the existence of mandatory state inspections, preferably accompanied by an effective, federally required warranty.[21]

Warranty and State Inspections

Objectives. It has been shown that the effectiveness of the recall provision—and indeed the effectiveness of the Clean Air Act in reducing air pollution—depends heavily on the incentives to car owners to perform the required maintenance. The incentives in turn depend on the existence of a mandatory state inspection procedure. In states without such inspection systems, car owners will be unlikely to perform any more maintenance than is normally necessary to keep their vehicles in good running condition. But keeping emissions control systems functioning properly (especially in projected 1976 models) requires special maintenance, which may even interfere with normal maintenance since good performance, driveability, and fuel economy are not compatible with good emissions performance. In short, without a state inspection, vehicles will not be subjected to the maintenance allowed by the federal certification test, and may therefore be characterized by emissions five to ten times the desired levels.

The administrator of the EPA will decide what states need inspection systems in order to meet their ambient air standards. If the implementation plan submitted by a state does not include a state inspection system, then the administrator must either require the state to revise its plan to include vehicle inspection or acknowledge that the state does not need vehicle inspection in order to meet the ambient standards. The latter option implies that, even if there is *no* vehicle maintenance beyond what is necessary to insure performance, driveability, and fuel economy, the emissions from motor vehicles will not cause ambient pollution levels in the state to exceed the standards. Since the emissions from a poorly maintained vehicle (or a vehicle in need of a catalyst replacement) may exceed normal maintenance emissions by a large factor, this assumption does not seem to be valid for most states.[22] Nevertheless, only a

---

[21]Under the Clean Air Act, of course, a recall may also be initiated by the EPA if there is a clear manufacturing defect that would cause excess emissions, testing notwithstanding.

[22]In general, the validity of the assumption would depend on factors such as car density, traffic patterns, number and size of urban centers, mass transit availability, employment distribution, urban topography, and general meteorological conditions.

handful of states included state inspections in their 1972 Implementation Plans.[23]

The purpose of the federally required emissions warranty is to protect the car owner who performs the required maintenance from suffering state penalties and excessive repair bills. According to Section 207(d) of the Clean Air Act, this burden must be borne by the manufacturer, and may not be transferred to the dealer. By shifting the cost of failure from the car owner, the warranty reinforces the incentive to perform the required maintenance, since only by performing this maintenance is the owner protected under the warranty. Furthermore, the car owner is relieved of any perceived need to perform excessive maintenance.

Also affecting the owner's decisions about maintenance will be the penalty, if any, for failing the state inspection; the corrective maintenance costs required if the inspection test is failed; the measurement error of the test; and the owner's perceived uncertainty about passing the test. Everything being equal, a stricter state standard and a noisy (high measurement error) test will induce the owner to perform more maintenance.

The nature of the state inspection test and criterion affects the manufacturers' incentives as well. First, even though the inspections are aimed at individual drivers, the results will, or should be, channeled into a recall process. Second, the strictness of the state standards and the variance of the test used will affect the manufacturers' choice of design parameters and therefore vehicle cost. The impact of these factors depends, of course, on the effectiveness of the warranty; in its absence, not only will individual inspections have no effect on manufacturers (except through good will), but the recall will be less potent because owners will have less of an incentive to perform adequate maintenance. State inspections with an effective warranty can fail to affect manufacturers' decisions only if (1) decision-makers in the auto industry today do not perceive the cost of action under the warranty in the future as falling on them, or (2) the threat of enforcement under the warranty is not credible to them, for legal or other reasons.

Performance. Refer back to the objectives set out earlier. To meet the primary objective of insuring adequate maintenance, we would like to have a high state penalty, an effective warranty, and a strict state standard. The third objective, however, comes into conflict here. A strict standard and an effective warranty, given nonzero testing and production variation, will force manufacturers to reduce the mean emissions substantially below the standards. In addition, the fourth and fifth objectives come into conflict with the second if the warranty provision is ineffective, because owners performing maintenance will

---

[23]The response of the EPA to these proposed plans in 1973 was to modify many of them to include state inspections. At this time, it is not clear whether or not the EPA will be able to make its rulings stick.

often fail the test because of process and measurement variance and the strict standard, and because cautious owners will spend too much on maintenance to avoid failing the test. If the warranty is effective, however, the fourth and fifth objectives pose no problems, since performing the manufacturer's required maintenance will sufficiently protect car owners. The trade-off is then between the need for a strict standard under the second objective and the need for a lax standard under the third.

It is shown in the Appendix that under the CVS Test, manufacturers select the emissions mean and variance, and the deterioration mean and variance, so that the marginal cost of improving each parameter is proportional to the marginal contribution of that parameter to reducing the probability of failure; that constant of proportionality is the "shadow price" of increasing the probability of failure, namely the cost of failure. By selecting the parameters, the manufacturers implicitly settle on a level of probability with which they may fail the CVS Test. We want to design the other averaging tests (e.g., assembly-line, recall) so that, given the same parameters, the probability of failure remains the same or slightly less.

Clearly, for *individual* inspections, the cost to the manufacturer under the warranty of an individual vehicle is not nearly so high as the cost of, say, a total recall. If N vehicles are produced, then the cost of each individual failure will be approximately $1/N$ times the cost of cancellation of the product line or total recall. This cost will tend to be somewhat greater because of loss of goodwill, but may be less than the cost of canceling the product line if the companies are making profits well above the cost of repairing all vehicles on the road. Assume that these two factors cancel each other, and attribute a cost $1/N$ times the cost of cancellation to each individual failure. To keep the manufacturers' expected losses constant for the same batch of vehicles, the probability of failure of the individual inspection test will have to be the same as the probability of failure of the CVS Test based on an average of, say, n vehicles.[24]

To achieve the same level of probability for individual tests as for certification, we must either make the inspection test *more* precise than the CVS Test (even allowing for correlation error) or raise the nominal standards. The former is technically impossible. The latter is the only possibility, but it may be politically infeasible to raise the standard on the road. At any rate, it is possible to compute the necessary adjustment in the nominal standards to order to keep the probability of failure constant for the prototype and on-the-road tests. The increase would be at least an order of magnitude under reasonable assumptions. Thus, instead of using 3.4 gm/mi as the CO standard, it might be necessary to use 17 gm/mi, or even 34 gm/mi or higher.

---

[24]This is the case if manufacturers are risk-neutral. If they are risk-averse, then the probability of failure for the individual test may be somewhat higher.

There may also be legal reasons for the nominal on-the-road standard to have to be substantially higher than the target emissions levels. In particular, if a vehicle fails a state test, it may be necessary to demonstrate "beyond a reasonable doubt" that the vehicle is, in fact, in violation of the actual standards. For example, a vehicle with true emissions below the actual standard might be required to have no more than a 5 percent or 1 percent chance of failing the test. This would be desirable both to protect the manufacturer under the warranty and to protect the car owner if no warranty existed.

Consider the CO example used earlier and in the Appendix. With a test that correlates as $r = .1$ with the CVS Test, and with the other sources of variation as given in the Appendix, the standard error of a single test is approximately ±4.5 gm/mi, compared to the standard of 3.4 gm/mi. Assuming a normal distribution,[25] the nominal standard at which the probability of failure for a car with true emissions of 3.4 gm/mi would be 5 percent is about 10.8 gm/mi; a level of 1 percent would require a nominal standard of about 13.9 gm/mi. If we relax the assumption that the correlation of the test with CVS is as bad as $r = .1$ and (at least for CO) assume that the correlation of an Idle Test or Key Mode Test with CVS is $r = .5$, then the standard error of a test falls to ±0.8 gm/mi, the standard that guarantees a 5 percent probability of error of commision is 4.7 gm/mi, and the standard that guarantees a 1 percent probability of error of commission is 5.3 gm/mi.

The problem with such a lax on-the-road standard is that owners who do *not* perform maintenance are likely to pass the test, in contradiction with the primary objective of on-the-road testing. If no nominal standard can be found that is high enough so as not to put an undesirable burden on manufacturers, but still low enough to induce car owners to perform reasonable maintenance, then the warranty system cannot be effectively implemented.

There is reason to believe, however, that cautious owners will perform the required maintenance under the high standard-*cum*-warranty system, provided that the state penalty is severe enough. (Otherwise, states could go ahead and set tighter standards, without a warranty, and the car owners would pay the price for negligence on the part of car manufacturers. This state of affairs could be improved somewhat if manufacturers were required to correct the highest emitters at the assembly line.)

On balance, we conclude the following: due to the enormous variability of *individual* tests (as opposed to averages), it is imperative that any *equitable* state standard enforced on an individual basis be higher than the federal target levels applied to averages. Such relaxed standards are still likely to be sufficient to induce car owners to perform maintenance under an effective warranty system.

---

[25] Obviously, due to the skewness built into the situation because emissions cannot be negative, the approximate normality is assumed to hold at least for the right tail.

The decision as to which state tests to authorize under the warranty system is critical. On balance, the need for good correlation with the federal CVS Test may, for the reasons given earlier, be less critical than the need for a robust testing package including different tests at each stage. A test such as the Key Mode, with its diagnostic value, may be more valuable for state inspections than a more expensive test modeled after CVS.

The use of a mix of tests to be applied at the three stages—prototype, assembly-line, on-the-road—is a sensible policy, despite the variability introduced as a result of imperfect correlation. Using, say, the CVS Test for prototypes, the Idle Test at the assembly line, and the Key Mode on the road, would prevent the manufacturer from designing cars to pass the test rather than to minimize pollution. The advantages of robustness outweigh the disadvantages of imperfect correlation.

Two important points need to be made before we turn to the costs and implementation of state inspection systems. The first point concerns emissions of vehicles *after* 50,000 miles. If, in fact, half of all miles driven are in cars with mileage above the 50,000-mile average, then what happens to the emissions of these cars is of vital importance in determining the performance of an on-the-road testing program. If these cars are not tested, then one may assume that all emissions-directed maintenance will stop and that the emissions of these vehicles will soar. This would be the case if, for example, a catalyst required replacement every 25,000 miles but was not replaced at 50,000 or 75,000 miles.

The conclusion to be drawn is that some form of inspection or mandatory maintenance must be applied to vehicles even after the warranty runs out. If emissions testing—rather than mandatory maintenance alone—is chosen, then clearly the nominal standards applied should be higher than the standards applied to under-50,000 mile vehicles because of built-in deterioration. In short, state inspection and maintenance regimes should not be limited to cars with mileage under 50,000 miles. The standards applied to over-50,000-mile cars should, however, be numerically higher to allow for built-in deterioration beyond 50,000 miles.

The second point concerns the visual inspections which must be included in state inspections to make them eligible for a federal subsidy. It is possible that visual inspections alone would do a great deal to reduce pollution simply by preventing the less sophisticated forms of tampering. There should also be strict enforcement of penalties for any individual who tampers with the emissions control devices of a vehicle.

Costs and Implementation. Given that on-the-road testing is absolutely essential to insure that car owners perform maintenance, and given that such maintenance is absolutely essential to insure that vehicle emissions are kept under reasonable control throughout the useful life of a car, the following questions arise: (1) what would on-the-road inspection cost? and (2) in any case, can on-the-road inspection be widely implemented? These and other related questions are addressed in the next chapter.

Appendix

Statistical Models Used in the Analysis

FORMULATION OF THE TESTING MODEL: PROTOTYPE STAGE

The model described here applies to the analysis of all three testing stages with some modifications. Consider a single engine type and a single pollutant. The same qualitative results would hold in the three-pollutant case, but the miltivariate statistics are tedious and do not add to an understanding of the problem at this time.

Assume that the 4,000-mile emissions of this particular vehicle model are distributed according to a normal distribution with mean μ and variance σ^2. The variance is essentially the variance in mass production. Assume further that the deterioration in emissions is linear, and that the *difference* between emissions at 4,000 miles and emissions at 50,000 miles is distributed normally with mean β and variance σ_δ^2, and independently of the level of emissions at 4,000 miles. These quantities are assumed to be based on maintenance according to the manufacturers' specifications. The variance in deterioration is due to production variance and to differences in maintenance procedures within the specification tolerances. Finally, consider the CVS Test, defined to be an unbiased estimate of true emissions, with measurement variance τ^2.

The steps in the modified CVS procedure using averages are as follows: Select m durability vehicles at random to get an estimate of the deterioration factor β; call this estimate $\hat{\beta}$. Then measure n test vehicles at 4,000 miles only. To each score, add $\hat{\beta}$ and take the average. This average is an estimate $(\mu \hat{+} \beta)$ of $\mu + \beta$ which is unbiased and has variance

$$v^2 = \frac{\sigma^2 + \tau^2}{n} + \frac{\sigma^2 + 2\tau^2}{m} . \tag{1}$$

Consider now some data for CO emissions. Based on production coefficients of variation of 20 percent at 1975 levels, we get

$$\sigma^2 = [(.2)(3.4)]^2 \approx 0.4 \quad (gm/mi)^2.$$

Based on 100 percent emissions deterioration and a 50 percent coefficient of variation about that mean, we get

$$\sigma_\delta^2 = [(.5)(1.7)]^2 \approx 0.75 \quad (gm/mi)^2.$$

Based on a 20 percent coefficient of variation for the CVS Test itself, we get, at 1975 CO levels,

$$\tau^2 = [(.2)(3.4)]^2 \approx 0.4 \quad (gm/mi)^2.$$

Therefore, if $m = 4$ durability vehicles and $n = 4$ test vehicles are used, the net variance would be

$$v^2 \approx 0.58 \, (gm/mi)^2,$$

and the standard error would be

$$v \approx \pm 0.75 \, gm/mi,$$

compared to the emissions standard of 3.4 gm/mi. If $m = 10$ and $n = 10$ vehicles are used, then

$$v \approx \pm 0.5 \, gm/mi.$$

If μ^* is the standard to be applied, then

$$\text{Prob}\,[(\mu \hat{+} \beta) > \mu^*] = 1 - \Phi\left\{\frac{\mu^* - (\mu + \beta)}{v}\right\},$$

where ϕ is the cumulative distribution function of the normal distribution. In the CO example,

$$\text{Prob}\,[(\mu \hat{+} \beta) > 3.4] = 1 - \Phi\left\{\frac{3.4 - (\mu + \beta)}{0.5}\right\}$$

if $m = 10$ and $n = 10$. This is the probability of failing the test, for given design parameters μ, β, σ^2, and σ_δ^2. To reduce this probability to about 2.5 percent would require, for fixed variances σ^2 and σ_δ^2, design values of μ and β such that

$$\mu + \beta = \mu^* - 2v$$
$$= 3.4 - 1.0$$
$$= 2.4 \quad \text{gm/mi},$$

or about 30 percent below the standard.

The larger m and n are—that is, the more vehicles go into the average the smaller is the variance. Thus, for any fixed small probability of failure α, the expected emissions, $\mu + \beta$, will be less far below the standard if m and n are larger (so that v is smaller). All of this leads to the result stated earlier in Chapter 5, that the more vehicles are averaged in for purposes of applying the standard, the less manufacturers will be required to overshoot the standards in order to pass the test with probability $1 - \alpha$.

MANUFACTURERS' RESPONSE TO UNCERTAINTY IN TESTING

Assume that manufacturers can select the parameters μ, β, σ^2, and σ_δ^2, subject to a cost function $c(\mu, \beta, \sigma^2, \sigma_\delta^2)$. Define the function

$$Q(\mu, \beta, \sigma^2, \sigma_\delta^2 \mid m, n, \tau) = 1 - \Phi \left\{ \frac{\mu^* - (\mu + \beta)}{v(m, n, \tau)} \right\}.$$

If manufacturers want to reduce the probability of failure to some level, α, then the ratio of the marginal cost of improving the parameter to the marginal decrease in the probability Q of failure must be the same for all parameters. Symbolically,

$$\frac{\partial C/\partial \mu}{\partial Q/\partial \mu} = \frac{\partial C/\partial \beta}{\partial Q/\partial \beta} = \frac{\partial C/\partial \sigma}{\partial Q/\partial \sigma} = \frac{\partial C/\partial \sigma_\delta}{\partial Q/\partial \sigma_\delta} = \lambda.$$

The value of λ is simply the shadow price of reducing the level of α. If the cost of failure (including goodwill loss), wasted resources, and everything else is L, then clearly $\lambda = L$, provided that the manufacturer is risk-neutral. For a single, narrowly-defined line of vehicles, risk-neutrality may be a fair assumption. If the manufacturer is risk-averse, then the design parameters he will select will be even smaller than the analysis would predict.

Consider the decision of how many vehicles to test, as durability vehicles and as emissions test vehicles. If the cost of each of the m durability tests is c_m, if the cost of each of the n emissions tests is c_n, and if the nominal standard μ^* is fixed, then we should choose m and n to minimize

$$mc_m + nc_n + c(\mu, \beta, \sigma^2, \sigma_\delta^2 \mid m, n, \tau)$$

plus any fixed costs of testing. Thus the optimal number of each type of test is such that the marginal cost of the test is equal to the marginal reduction in producer's cost due to increasing the number of tests by one. Under the assumption that μ^* is the socially optimal level of pollution, the marginal loss in pollution benefits cannot be as great as the marginal savings if we move up toward the standard. As we move up close to the standard, however, we should consider the loss of pollution benefits in the optimization.

If we could change the *nominal* standard μ^*, then the problem would change substantially. In that case, we would set a higher *nominal* standard and, with less expense on testing and automotive design, we could expect to meet the average *target* standard.

CORRELATION AND RESIDUAL VARIATION

We will show that the notion of statistical correlation, in the usual sense, is less important than the notion of *residual variation* in a test after its score has been adjusted to correspond to the CVS score. It will then be clear what kinds of tests are desirable for use in conjunction with some reference test (such as CVS).

The score of the CVS Test is defined as

$$f = e + u,$$

where e is the true emissions level for the vehicles (for CO, say) and is drawn from a normal distribution of vehicles with mean emissions μ and variance σ^2. Assume that the error is independent of the level, e, and is distributed normally with mean 0 and variance σ^2. In other words, we define true emissions e on the basis of this particular test. Now consider a second test with a score g that will be a function of true emissions e and some random component. We will consider the case for which that function is linear, so that

$$g = be + a + w,$$

where a and b are constants[1] (characteristics of the test), and w is an additive error with zero mean and variance σ_w^2. Notice that while the error u includes only measurement error, the error w includes both measurement error *and* process error in the relationship between true emissions and emissions as

---

[1] It is assumed here and in the following analysis that a and b are known with certainty. In fact, they are probably derived from a regression analysis and are estimates $\hat{a}$ and $\hat{b}$ subject to error. If the sample size on which the estimates are based is large enough, however, then this error will be swamped by the process error w.

measured in the test g. That is, two cars with the same emissions e might, even if there were zero measurement error, come out with different scores g. We have defined e, however, so that the same could not be true for the measurement f.

It can be shown that f is an unbiased estimate of emissions e, with variance τ^2, and that $(g-a)/b$ is also an unbiased estimate of e, with variance σ_w^2/b^2. The interpretation of $(g-a)/b$ is that it gives an *adjusted score* on the second test which can be compared to the score on the first test. If b is zero, then the two tests are absolutely uncorrelated, and we cannot use the second test. If b is small but positive, the residual variance σ_w^2/b^2 will be manageably small if σ_w^2 is reasonably small. If b is actually negative, so that a high score on the first test corresponds to a low score on the second test, then it may be politically impossible to use the adjusted score of the second test for enforcement even if the correlation is perfect (-1).

Below are the formulas for the statistical correlation between the two tests in this model. They show that a low correlation necessarily implies a high residual variance, σ_w^2/b^2, and conversely. The results are

$$E[f] = E[e] = \mu,$$

$$E[g] = E[be+a] = b\mu + a,$$

$$E[(g-a)/b] = E[e] = \mu,$$

$$\text{Var}[f] = \text{Var}[e] + \text{Var}[u] = \sigma^2 + \tau^2,$$

$$\text{Var}[g] = b^2 \text{Var}[e] + \text{Var}[w] = b^2\sigma^2 + \sigma_w^2,$$

$$\text{Var}[(g-a)/b] = \sigma^2 + \sigma_w^2/b^2,$$

$$\text{Cov}[f,g] = E[([e-\mu] + u)(b[e-\mu] + w)]$$

$$= b\sigma^2,$$

$$\text{Corr}[f,g] = \frac{b\sigma^2}{(\sigma^2 + \tau^2)^{1/2}(b^2\sigma^2 + \sigma_w^2)^{1/2}}$$

$$= \frac{\sigma^2}{(\sigma^2 + \tau^2)^{1/2}(\sigma^2 + \sigma_w^2/b^2)^{1/2}}$$

Thus, we compute the residual variation in test g, σ_w^2/b^2, from the observed correlation with the CVS Test f, as

$$\sigma_w^2/b^2 + \frac{\sigma^2}{(\sigma^2 + \tau^2)(\text{Corr}[f,g])^2} - \sigma^2. \qquad (2)$$

This method is used in the following discussion to estimate the variation of a test observed to correlate poorly ($r = 0.1$) with the CVS Test.

MODEL OF ASSEMBLY-LINE TESTING

The testing criterion used in this model of assembly-line testing is essentially the same as that used in the model of prototype testing discussed above. A sample of m durability vehicles is tested at the prototype stage to obtain an estimate, $\hat{\beta}$, of the average deterioration β between 4,000 miles and 50,000 miles. A sample of n_a assembly-line vehicles is then tested, and the results are then adjusted to correspond to CVS scores in the manner described in the above discussion of correlation. (In practice, it would probably be simpler to adjust the nominal standard.) These scores are then adjusted for any bias due to green engine testing.

The variance of a single assembly-line measurement is therefore

$$\sigma^2 + \sigma_g^2 + \sigma_o^2 + \tau_a^2,$$

where σ^2 is production variance; σ_g^2 is the variance due to green engines; σ_o^2 is the residual process variance introduced by adjusting the score of, say, an Idle Test to correspond to CVS Test scores; and τ_a^2 is the measurement error. The variance of an average of n_a is therefore

$$\frac{\sigma^2 + \sigma_g^2 + \sigma_o^2 + \tau_a^2}{n_a}$$

We have assumed production variance for CO at 1975 levels to be $\sigma^2 = 0.4$ (gm/mi)^2 as derived above. We further assume that the green engine variance is approximately the same, as manufacturers claim, so that $\sigma^2 = 0.4$ (gm/mi)^2. The biggest contribution to the net variance, if a test such as the Idle Test is used, is the variation due to poor correlation with the CVS Test. By this method described above, we compute the value of $\sigma_o^2 + \tau_a^2$ corresponding to a (poor) correlation of $r = .1$. We obtain from equation (2) that

$$\sigma_o^2 + \tau_a^2 = \frac{(.4)^2}{(.4 + .4)(.1)} - (.4)^2 = 16/.8 = 20 \text{ (gm/mi)}^2.$$

An average of $n_a = 400$ therefore has variance of

$$\frac{20.8}{400} \cong .052 \text{ (gm/mi)}^2.$$

Clearly, if n_a were considerably larger—as it would be if 100 percent assembly-line testing were used—then the variance of an average of zero-mile emissions would be quite small.

The variance due to the extrapolation to 50,000 miles is, as before,

$$\frac{\sigma^2 + 2\tau^2}{m},$$

which has a value of 0.16 (gm/mi)^2 using the numbers that are used above. Thus, the total variance is $v^2 = .05 + .16 = .21$, so that the standard error is $v = 0.45$—smaller than the standard error of the hypothetical prototype test discussed above despite the poor correlation between tests. By the criterion that the uncertainty of the assembly-line test must be no greater than the uncertainty of the prototype test, an assembly-line test using averages is feasible, even if between-test correlations are rather poor.

In general, the prototype variance and the assembly-line variance are equal if

$$\frac{\sigma^2 + \tau^2}{n} + \frac{\sigma_\delta^2 + 2\tau^2}{m} = \frac{\sigma^2 + \sigma_g^2 + \sigma_o^2 + \tau_a^2}{n_a} + \frac{\sigma_\delta^2 + 2\tau^2}{m},$$

so that

$$n_a = n \frac{\sigma^2 + \sigma_g^2 + \sigma_o^2 + \tau_a^2}{\sigma^2 + \tau^2}.$$

We might ask under what conditions it is possible to achieve this level of confidence using 100 percent testing at the assembly line with, say, an Idle Test. Such a condition is that if k percent of the total production line is tested at certification, then

$$\frac{\sigma^2 + \sigma_g^2 + \sigma_o^2 + \tau_a^2}{\sigma^2 + \tau^2} = 100/k.$$

Unless $\sigma_g^2 + \sigma_o^2 + \tau_a^2$ is enormous, and considering that k is rather small, this will be the case for any reasonable test. When more data become available, it will be possible to verify this result.

Chapter Six

Testing and Maintenance of In-Use Vehicles

Jack M. Appleman

The preceding four chapters have illuminated areas where our present policy toward automobile emissions either simplifies or ignores critical aspects of the regulatory problem the government has undertaken. Chapters 2, 3, and 4 explored the likely outcome of attempts to meet the Clean Air Act's 1975-76 emissions deadlines if the catalyst-controlled ICE remains the basic propulsion system. The conclusion reached was that the 90 percent emissions reductions sought in the Act are unlikely to be achieved given the problems of deterioration inherent in the conventional ICE. Under a wide range of assumptions about cost and system deterioration, it appeared far better to shift to an engine technology less prone to degradation.

Chapter 5 then turned to the question of vehicle testing. The Act provides for prototype, assembly-line, and in-use testing as well as warranties and recalls in order to enforce efficient vehicle design and proper user maintenance. Chapter 5 analyzed how these provisions might interact, and three points in that discussion are relevant to the present analysis of in-use testing as a means for enforcing on-the-road emissions limits. First, the de facto standards are quite sensitive to such things as species definition, instrumentation, and aggregation techniques. Second, only a narrow band of test parameter settings actually leads to the desired result in terms of vehicle emissions and costs. If test conditions such as aggregation rules, test fleet size, and driving cycles result in low de facto standards, then automobiles may be overdesigned and unnecessarily expensive. If the tests are not properly specified, the inspection sequence will lack robustness and may result in excessive on-the-road emissions. Lastly, the previous chapter makes the point that the three stages of testing and enforcement are highly interdependent and must be carefully planned if an efficient scheme of regulation is to be achieved. In particular, most of the incentives for maintenance and design, as well as the stringency of the de facto standards, can be heavily influenced by the creation of viable state inspection and testing programs.

In this chapter we consider whether or not the EPA and the states should advocate such in-use testing. Both the previous chapter and the Act itself promise rich benefits from in-use testing, whereas this discussion focuses on the barriers that have to be surmounted in order to make state inspection a reality. The next section describes the states' incentives to adopt testing and the environment from which the manufacturers view in-use testing. The chapter also explores the likely effects on aggregate emissions of testing-induced maintenance, and thus attempts an evaluation of the ultimate motivation behind the whole road testing idea.

STATE IMPLEMENTATION OF ENFORCEMENT PROGRAMS

Under the Clean Air Act, each of the fifty states is given responsibility for any in-use inspection and maintenance programs that may be needed to insure compliance. The 1970 Amendments authorize the federal Environmental Protection Agency (EPA) to set ambient air quality standards for the nation as a whole, and require the states to formulate plans for achieving these standards. The states are given considerable latitude to decide the distribution of the requisite reductions among polluters, but state implementation plans are not to be approved unless they provide "to the extent necessary and practicable, for periodic inspection and testing of motor vehicles to enforce compliance with applicable emissions standards" [United States Code 1970]. The EPA has partially funded efforts to formulate these implementation plans, and the agency is authorized to pay up to two-thirds of the cost for any vehicle inspection and testing program that is actually adopted.

In the Act, the provisions for manufacturer's performance warranty and federal recall of vehicles are tightly linked to in-use testing by the states. The EPA administrator must rely on the results of state testing programs (or of voluntary surveillance) in any determination that vehicles are in noncompliance and must be recalled. The federal government may not institute a mandatory inspection program for this purpose. Likewise, the provision of the Act requiring a warranty of control system performance for 50,000 miles or five years cannot be implemented without state action. In order to qualify for service at the manufacturer's expense, the car must have been properly maintained, failed in a state inspection program, and the motorist subjected to some form of penalty for being caught in this unfortunate position. The choice of a road test is not totally at the state's discretion: the EPA must determine that the test scores correlate with the results of the federal certification test. Thus the language of the Act seems to foresee a development wherein the EPA and the states search together for a suitable road test, and the EPA helps to fund in-use testing where it is needed to help meet ambient air quality standards.

In fact, in the early months after the Act was passed, the EPA moved to discourage state environmental agencies from moving too quickly in advocating in-use inspection. In August 1971 the EPA promulgated a rule stating that only "at such time as [they are determined by] the Administrator to be feasible and practicable," programs for periodic inspection and testing of motor vehicle emission control systems will be considered acceptable control strategies [*Federal Register 36* (August 14) 1971, p. 15487]. Ostensibly, this cautious approach to inspection was adopted because the EPA had several studies under way to assess the effectiveness of alternative inspection schemes. The results of these studies were not expected before January 1972, the filing deadline for the state implementation plans. Consequently, states were advised that their 1972 plans needed only to identify tentative transportation control measures being considered, but that by spring 1973 they must have definitive transportation control plans.

The EPA studies, released in the spring and fall of 1972 [Environmental Protection Agency 1972], are not likely to resolve the question of in-use inspection. Although the studies found that inspections reduce emissions, the reduction is not extraordinary. Naturally, the only vehicles available for use in these studies were cars from the 1972 and previous model years. For precontrolled cars (pre-1968 models) and for cars with the first generation of emission controls (1968 to 1972 models), it was found that the maintenance of the high emitters identified in an inspection network could reduce overall emissions, though the reduction was not impressive.[1] Furthermore, it is very difficult to draw any conclusions from these studies regarding the likely effectivness of similar schemes when applied to cars designed to meet the 1975 and 1976 standards. The findings did little more than point up the uncertainty regarding the benefits to be realized from state enforcement activities.

Determinants of Success of State Programs

In January 1972, twenty states indicated that in-use testing was under consideration as an element of their air pollution control strategies. The decisions by the remaining thirty states not to propose road tests can be attributed to a number of factors including favorable environmental conditions, lack of EPA support for the idea, or a concern that in-use testing may be unworkable. On this last point there is real cause for concern. The likelihood of successful implementation of such schemes on the state level seems to depend on three factors—financial cost, organizational changes, and public reaction. A close look at each of these aspects makes it appear that a prodigious effort will be required to make state enforcement schemes work smoothly and effectively.

Financial Cost to State Governments. Direct governmental cost is probably the most predictable, controllable, and salient feature of in-use inspection programs, and thus it tends to dominate discussion of the issue. Depending

---

[1] For an annual inspection failing about 30 percent of the cars each year and sending them for emissions control service, only about a 10 to 12 percent reduction was achieved for the total vehicle fleet [Environmental Protection Agency 1972].

on the test technology chosen, EPA estimates of operating costs run from $1 to $5 per car, with the associated capital costs varying from $10,000 to $1,000,000 per inspection lane [Northrop Corporation 1971, p. 6-2; and New York City Environmental Protection Agency 1972, Section 7, p. 2]. The number of lanes needed is determined by the volume of cars subject to inspection, the duration and efficiency of testing, and the number of retests required. A study prepared for the State of California by the Northrop Corporation [1971, p.8-1] concludes that capital costs for state-operated testing of California's 12 million cars would be between $10 million and $88 million, depending on the tests selected. In its Regional Air Pollution Plan the New York City Environmental Protection Agency [1972] estimates that a statewide inspection plan for New York would require a $15 million capital investment. A capital expenditure of $5 million would be required if the program were restricted to New York City alone.

Apart from investment in physical facilities, the start-up costs of an inspection program are likely to be substantial. The Northrop study suggests that it would take two years to implement a state-owned and -operated program in California. In this period the costs of planning, situating, and constructing inspection stations—along with training personnel, educating the public and the repair industry, and developing information systems—would be a major expense, at least equal to a year's operating budget. Undefrayed by user charges and possibly only partially reduced by the federal subsidy provided for in the Act, start-up costs could be $5 to $10 million in the more populous states.

Estimates of this magnitude provide significant incentives to seek cost-reducing features, and two alternatives to state-owned and -operated inspection stations are receiving attention. One is to combine state-run emissions tests with existing safety inspection. However, since only New Jersey, Delaware, and Washington, D.C., have state-owned and -operated safety inspection lanes, this option is not widely available at present. Only in New Jersey has such a system been found to be feasible without federal funding. The second alternative is to authorize private garages and gas stations to perform the emissions inspections. While this approach might lower direct state budgetary costs, it might also reduce the amount of federal funding available. In addition, turning to private enterprise to reduce public cost may make it more difficult to gain public acceptance of the scheme. Quality control would be difficult to enforce, and an elaborate certification procedure might be required just to convince motorists they were receiving fair treatment at the hands of profit-oriented local garages.

Organizational Requirements. A state must assess its ability to provide the people and the organization to manage and operate a testing program. The lack of trained labor for testing may be critical. The Northrop Study indicates that the 400 lanes necessary to blanket California will require 400 technicians with one to three years' experience, 400 inspectors with three to ten years' experience, and 100 station managers with ten or more years' experience. Additionally, the study suggests that all personnel should be required to participate in a specialized training program and pass a certification test. Today these skilled personnel simply are not available.

Bureaucratic impediments and resistance may also be difficult to overcome since inspection may impinge upon the organizational prerogatives of the highway department or the state motor vehicle agency. The reorganization of motor vehicle regulation in an effort to ensure the efficiency of emissions inspection may entail radical changes in the mode of operations of existing state agencies.

These difficulties of personnel recruitment and governmental reorganization are particularly troublesome in the light of the possible early obsolescence of the program. The state enforcement system is needed only if motor vehicles are prone to deterioration, and if this deterioration can be reversed or retarded by maintenance. The catalyst-controlled vehicles expected in the mid-1970s possess these attributes, but in the event a stable low-emissions vehicle is developed and adopted the whole organizational and financial burden would have been undertaken for nothing. Since the costs are high and the lead time in setting up inspection programs is so long, states are likely to be hesitant to gamble on the outcome of current dispute about automotive propulsion technology.

Public Acceptability. An inspection system can fail to attain public acceptability in either the testing or repair phases or both. Test credibility may be difficult to establish for two reasons: motorists will not be able to detect most failure conditions without a test, and the test results are not always repeatable. The situation is much more difficult than that faced in safety inspections. For example, any motorist can tell if his horn is silent or his brake lights are out, but it's very hard to know if your exhaust gas is recirculating, or if you have overly high levels of odorless, tasteless emissions. The National Academy study reports that "there is a significant variation in results when the same vehicle is tested several times . . . variations in emissions of 50 percent [above and below] the average value are not uncommon" [National Academy of Sciences 1972, p. 22].

Even assuming that the public knows of the variations in emissions testing, how many retests does one perform before convincing a motorist that his new car fails the emissions standard while older cars may pass? The situation can get very awkward if the motorist faces a repair bill of $20 to $30 or loss of his car registration, yet can't see that anything is wrong. While 75 percent of car owners in a sample survey conducted in California "[believed] a mandatory vehicle emission inspection and corrective maintenance program [was] necessary" [Northrop Corporation 1971, p. 3-1] they also may have been thinking it would be the "other guy" who would fail the inspection. One solution to the credibility problem is to improve the test technology through a form of diagnostic testing that isolates the malfunction causing excess emissions. Unfortunately, the more sophisticated tests involve higher capital and labor costs.

In addition to establishing the credibility of testing, there is the problem of engendering public confidence in repairs. The Northrop study revealed a "general lack of motivation on the part of the mechanics who performed service on the test vehicles" [Northrop Corporation 1971, p. 11-5]. For both the California and New Jersey studies, garages were specially selected, and the managers realized the experimental implications of their service. Nevertheless, "the majority of garages performed . . . without a real interest. . . . Quality control was generally lacking [and] though . . . fraudulent practices were not detected . . . a number of services were of suspect quality" [Northrop Corporation 1971, p. 11-5].

Similar results were achieved in a study prepared for the state of New Jersey which analyzed performance by dealerships, maintenance service centers, and gas stations [Andreatch, Elston, and Lahey 1971]. It was found that even though repair service quality improved over time, a discrepancy remained between laboratory and private cost of repair. The study calculated that total repair cost should range from $8 to $17 depending upon the emission problem, but found in actual practice that private service costs were between $18 and $25. Excess replacements were made by the private service industry on almost every category of parts. It may take only a few publicized incidents of gross overcharge or subsequently failed tests to give a significant segment of the public the feeling that they are being cheated.

Furthermore, it is not clear that the cost is acceptable, even if the work is done properly. The Northrop survey of car owners, which showed that 75 percent favored mandatory inspection, found that only 50 percent thought a maximum of $10 was a reasonable repair bill. The Clean Air Amendments of 1970 appear designed to avoid consumer resistance on this score by requiring manufacturers to warrant parts and performance. Unfortunately, as will be argued below, circumstances will probably conspire to force many owners to have emissions service done at their own expense. Finally, even with prepayment through the manufacturer's warranty, public acceptability of in-use inspection may be adversely affected by the impact of pollution control systems on auto performance—particularly gasoline mileage, acceleration, and starting. Deliberate tampering with pollution control devices by owners seems as likely as voluntary repair.

In summary, in-use inspection faces formidable obstacles on all three decision criteria: cost, organizational requirements, and public acceptability. Of course, if a state has a significant pollution problem the choice will not merely be a "yes" or "no" to inspection; it will be between accepting inspection or another control strategy. EPA has suggested that states consider fuel conversion, auxiliary control devices, taxes, gasoline rationing, parking restrictions, staggered work hours, and mass transit. Of course, none of these alternatives is without difficulties.

Experience with State Testing Schemes

This pessimistic assessment of the likely future of state inspection systems is consistent with the experience of states that have tried to institute such schemes or have thought about it. In 1966, the New Jersey Pollution Control Act was amended to authorize the department of Environmental Protection to promulgate the standards and requirements for vehicle inspection. Between 1966 and 1972 the department instituted wide-ranging tests, surveys, and experiments in anticipation of being allowed to add emissions inspection to the repertoire of the state's thirty-four safety inspection stations. Finally in April 1971, on the politically propitious "Earth Day," Governor Cahill announced that his administration intended to implement compulsory testing and compliance. During the summer, while the inspection standards were being drafted and presented at public hearings, powerful opposition formed within the state administration. Led by the past and present Motor Vehicle directors, the Labor and Industry commissioner, the director of Consumer Protection, the state treasurer, and the budget director, the opponents argued that the service industry would be unable to handle the repairs. Unscrupulous mechanics would cheat drivers, who would in turn blame the state government. The proposal was shelved, ostensibly due to cost and test time. Finally in December the state announced that emissions inspection (with a simpler test than was originally proposed) would become a standard part of the vehicle inspection system, but enforcement of standards would be delayed for one or two years.[2]

Implementation plans prepared for the EPA by Massachusetts and New York indicate the real difficulty of formulating a coherent implementation plan. New York has two Clean Air Act implementation plans, one provided by the state [New York State Department of Environmental Conservation 1972] and the other by the New York City Environmental Protection Agency [1972]. The state plan proposes a spot-check inspection program; the city proposal claims that a more expensive and complex diagnostic test "will be the only truly meaningful and cost-effective approach" [New York City EPA 1972, p. 6]. In Massachusetts the Bureau of Air Quality Control submitted an implementation plan calling for parking restrictions and a moratorium on public garage construction two weeks after the Department of Public Works and the city of Boston launched a $300,000 study to develop a master plan for metropolitan parking [*Boston Globe,* January 6, 1972].

Thus, although inspection has found its way into the implementation plans of some twenty states (perhaps because of the tentative nature of those plans), those states seriously considering in-use inspection are deeply immersed in the problems just discussed. The federal EPA is in a position to influence state decisions, particularly through its control of funding grants. But whether EPA will encourage adoption of in-use inspection is still an open question. As this discussion suggests, it may take large amounts of money, time, and organizational skill to start up an inspection scheme. Yet the results in terms of improved maintenance and reduced emissions remain uncertain.

---

[2] A review of this experience was provided by the *Trenton Evening Times,* January 5, 1972, p. 1.

MANUFACTURERS' RESPONSE TO WARRANTY AND RECALL

As pointed out in Chapter 5, getting owners to perform adequate maintenance is not the only reason for adopting a state inspection system. The warranty and recall provisions of the Act, if called into action, could give manufacturers a continuing incentive to improve vehicle design in the areas relating to emissions control. Before drawing too pessimistic a conclusion about these state programs, it is well to think about their possible indirect influence on vehicle design.

The automobile manufacturers have vigorously maintained that it is not feasible to implement the warranty provisions of the Clean Air Act. In August 1970, Ford Motor Company's presentation to a Senate subcommittee emphasized that "a performance warranty rather than a [warranty against] failure of a specific piece of hardware . . . is completely unenforceable and impractical" [U.S. Senate, Committee on Public Works 1970, p. 1605]. Ford supported its conclusion by arguing that dealers would have neither the testing nor engineering capabilities necessary for repair, and that, in addition, it would be impossible to determine if a vehicle had been "properly maintained, serviced, and operated" [p. 1605]. The Automobile Manufacturer's Association testified that the five-year/50,000-mile warranty put an "impossible burden" on manufacturers, arguing that if pre-certification and pre-sale tests do not correctly predict lifetime performance, then the solution is better pre-testing [p. 1579]. Even officials at EPA seem to consider that warranty work is primarily a manufacturer-consumer problem with little likelihood of enforcement.

Although recall provisions seem to pose a greater threat to manufacturers than warranty work made necessary by in-use inspection, recall suffers from many of the same problems. Recall has to be based on testing of a "substantial number" of any class or category of vehicles or engines, yet the EPA is limited to voluntary surveillance or to state-provided inspection. And thus the EPA may have great difficulty marshalling the data to support a recall action. Furthermore, even if a recall action were taken, the per-car cost may not be any greater than that of warranty repairs because the EPA must still negotiate a plan with the manufacturers for remedying the nonconformity. Such negotiations may allow manufacturers to argue their way out of repairing some cars, or certain parts of the control device, that might have to be repaired if brought in by an angry customer who has failed in-use inspection.

Previous Experience with Warranty and Recall. The automobile manufacturers have experimented with various warranty plans and are presently subject to government-initiated recalls under the 1966 National Highway Safety Act. During the first sixteen months of the Safety Act, over 4.5 million vehicles of various model years were recalled, 65 percent for steering defects. These recalls cost the industry $50 to $100 million [Federal Trade Commission, pp. 154, 170]. Yet recalls continue today. The significance of this experience is open to interpretation. Perhaps the recall campaigns are not a very good incentive for meticulous design and careful manufacture. Organizational and cost considerations may make it easier for the manufacturers to detect and correct defects via recall rather than through testing and assembly-line inspection. It also could be that safety recalls are just too few and too limited to elicit a stronger response from the design laboratories. If so, large-scale emissions recalls might raise the costs to the point where manufacturers would respond.

Between 1961 and 1969, Chrysler, Ford, General Motors, and American Motors made warranty provisions a significant element of their competitive strategy. From a ninety-day/4,000-mile warranty in 1961, manufacturers moved to a five-year/50,000-mile power train warranty in 1967. Since 1969 there has been a substantial decline in the length and coverage to only one-year/12,000-mile warranties. The decline might have been due to unexpectedly poor quality control. Or it may be that once the competitive extreme of a five-year/50,000-mile warranty was reached, it was possible for everyone to cut back.

A Federal Trade Commission (FTC) staff report [1968] provides evidence for a different conclusion: the auto companies were unable (or unwilling) to manipulate their dealer organizations toward successful implementation of a comprehensive warranty plan. The FTC presents an impressive array of observations to support conclusions regarding three main causes for the unsuccessful warranty programs. First, owners do not read or understand the warranty provisions. As a result, dealers are often able to abuse these provisions. Judicial remedies are often unknown to the owner or are just too expensive. On the other hand, many customers make unjustified warranty claims and complain when repairs are not made.

Second, many dealers believe that they receive inadequate compensation from manufacturers for warranty-related work. Low compensation for predelivery inspection results in cursory inspection, with dealers often depending on the warranty to catch defects. On warranty work, manufacturers are treated as wholesale customers; they pay dealers for parts at cost plus 25 percent (versus the normal charge of cost plus 42 percent), and the flat rate and time allowances are such that reimbursement for warranty repair is below that for ordinary repairs. The manufacturers also disregard inventory costs and diagnostic time, and compensation is denied for certain disputed claims (perhaps as high as 2 percent of the total warranty volume). All these are disincentives for the dealer to perform, or perform well, warranty-financed repairs.

Third, the incentives directed at dealers by the manufacturers are not designed to ensure fulfillment of warranty obligations. The system of "performance competitions" among dealership regions is based, in part, upon underspending the warranty budgets. Manufacturers provide advice on adequate service facilities, but there is no evidence that they enforce their recommendations. In those aspects where sales emphasis can be compared to service emphasis (e.g., training facilities, accountability), sales are still foremost.

The FTC study concluded that "although the manufacturers have a firm control over the character of dealer operation . . . they have not as yet taken steps to make the performance of warranty repairs live up to the standards implied by the written warranty" [1968, pp. 197-198]. Thus it may be foolhardy to rely on the manufacturers to take the steps necessary to live up to the warranty that Congress has written into the 1970 Clean Air Act.

Effect of Warranty Experience on Vehicle Design. One argument in favor of a heavy emphasis on making the warranty and recall provisions work is that the warranty experience may reverberate back to the design laboratories. If cars are failing on the road, the vehicle design will be changed in order to avoid the service costs. If automobile manufacturers can be viewed as unitary, rational entities, as they have been described in Chapter 5, then it is likely that the warranty and recall provisions would have this effect. However, a close look at the design and production processes followed by automobile manufacturers indicates that the feedback of service experience on vehicle design cannot be expected to occur quickly. The design, manufacturing, and marketing process is so decentralized and lengthy that the warranty and recall costs are likely to have only a remote and insubstantial effect on the critical production decisions.

Take General Motors as an example. The firm consists of 35 manufacturing divisions, a central staff, and miscellaneous special facilities and subsidiaries. The corporation's operating philosophy is "decentralization with coordinated control" [Learned 1961, p. 1]. Each division is under the direction of a general manager who possesses a large degree of operating discretion within his own division. Automobiles are designed, developed, manufactured, and merchandised by separate divisions. When coordination between divisions becomes necessary, it is accomplished by two governing committees and three major staffs (operations, legal, and financial). The operations staff includes the engineering group, which plays a major role in pollution matters.

Although evidence is limited, it appears that GM's pollution control decisions result from a complex and lengthy process. Initially, scientists within GM's research laboratories conduct studies of possible control systems and their engine performance and emission characteristics. These studies are then submitted to the engineering staff, which is responsible for developing several alternative pollution control systems that will fit the emission-controlled engine into the total automobile design. The resulting alternative system designs are sent to a staff policy group which selects one or two preferred systems on the basis of emission levels, costs, and performance trade-offs. Performance measures probably dominate emission considerations at this stage. Next, division managers and chief engineers tailor the emissions system to provide enough leeway to allow for coverage of variances necessarily present in the assembly-line process (see Chapter 5) and still pass an assembly-line emissions test. Finally, the design is considered by the division sales and distribution manager.

It is this sales manager who will feel the consequences of massive warranty and recall repairs should they occur. But on a day-to-day basis, he has other more immediate and, from his perspective, more important problems to deal with. His primary task is to manage advertising, dealerships, list prices, and discount schedules so as to maximize sales. Problems that may be in the offing one, two, or five years in the future will be of lesser concern. Even if future sales were hurt by a past model's poor pollution control design, it is possible that the problem, hidden among many other factors, would not be clearly recognized.

The sheer length of the process from design to sales and from sales to in-use inspection also reduces the value of the warranty/recall provisions of the Act as an effective incentive for manufacturers in their design decisions. The GM divisions start planning for new model vehicles two to three years in advance of the fall unveiling. For example, division production decisions for the 1975 Chevrolet, which will be unveiled in fall 1974, began in spring 1972; and the major mass production features have been fixed since the beginning of 1973. Past production schedules indicate that the automobile will undergo prototype testing and certification during 1973. As suggested below, testing at the design stage can never fully duplicate the strains which the system will undergo in actual use. If the emissions system fails after sale to the customer, the detection by in-use inspection and subsequent recall is from three to seven years removed from the faulty design decisions. Even if GM responds promptly in redesigning its new cars after detection of a design flaw, it may require another two to three years for this redesigned system to reach the road.

In summary, the Clean Air Act assumes a corporate consciousness and design flexibilily that probably is not present. Perhaps emissions control will someday reach the level of emphasis presently given to marketability factors. In the meantime we must deal with the fact that warranty and recall represent incentives that are removed from industry decision-makers in terms of personal responsibility, outlook, and time horizon. Furthermore, as far as affecting design is concerned, the impact of an operating inspection system should not be expected before the end of the decade. In short, the Act's pre-sale certification requirements (perhaps supported by the revised structure of fines suggested in Chapter 4) may present the only effective "clubs" to bring about design changes.

A MODEL OF AUTO EMISSIONS INSPECTION AND ENFORCEMENT

The preceding section makes it clear how difficult it would be to establish an effective set of state programs of emissions testing and enforced vehicle maintenance. But the fact that it would be difficult to accomplish does not mean that on-the-road enforcement should not be attempted. The judgment turns on estimates of how badly emissions control systems may deteriorate on the road without periodic maintenance, and on how effective state inspection and maintenance systems might be in correcting high emitters. To investigate this issue we have developed an empirical model that can allow testing of alternative assumptions about these unknown characteristics of the situation to be faced in the late 1970s and beyond.[3] The

---

[3]The model shown here was developed jointly by the author and Henry D. Jacoby. Special thanks are due to Thomas Shemo for valuable assistance with computer programming.

model was described briefly in Chapter 3 where some of the model results were used in the analysis of broad policy options. At this point we can describe its structure in greater detail and look a little more closely at what the analysis reveals about state inspection schemes.

The Deterioration of Emissions Controls

In Chapter 5, the phenomenon of deterioration of emissions control system performance was discussed briefly. The key concept introduced at that point was that of "smoothed emissions deterioration" under a regime of adequate or "normal" maintenance. As shown in Figure 5-5, the smoothed emission rate for a typical vehicle is the result of two different factors. First, there is the normal wear and tear and aging of the vehicle which would yield some deterioration with the buildup of mileage, m, even with continuous, perfect maintenance. Emissions under these hypothetical conditions are indicated by the dashed function in Figure 5-5. To this must be added the "between maintenance" deterioration, which yields a more rapid rise in emissions with mileage, but which is reduced each time maintenance is performed. The combined result of normal aging and between-maintenance deterioration is the saw-toothed line in the figure, and this may be smoothed to produce the emissions function $e(m)$. Figure 5-5 portrays $e(m)$ as a linear function—an assumption also made in the models of deterioration that lie behind the federal emissions tests. In the federal certification procedure, the slope of the smoothed deterioration function is determined from a small sample of cars (see Chapter 5)—one function being estimated for each model and engine type. In order to pass certification, the test result at 4,000 miles, adjusted to 50,000 miles using the deterioration function, must be below the legislated standard.

Unfortunately, the data developed by this procedure and illustrated in Figure 5-5 do not give an adequate indication of the emissions that can actually be expected from controlled vehicles on the road. Emissions are almost certainly going to be higher than indicated by the certification test. Earlier in the book, several reasons for this outcome were suggested.

1. *Test-to-Road Deviation.* At the prototype stage of production, it is not feasible to conduct a test of emissions control durability which subjects cars to the operating conditions they will encounter in actual use. The problem is not the federal emissions measurement procedure itself or even the methods for calculating deterioration factors from data on 50,000 mile durability vehicles.[4] The difficulty arises in the conditions to which prototype cars are subjected during the accumulation of mileage for the durability test. Basically, what happens now is that a fleet of prototype vehicles is driven around a test track twenty-four hours a day at 30 to 35 mph for as long

---

[4]See Chapter 5 for the details of the procedure. Chapter 5 discusses several changes in the federal test (the CVS Test, as it is known); their adoption or rejection is not important to the problem of durability, which is our concern here.

as it takes to run up 50,000 miles. The mileage is accumulated at the manufacturer's facility by his own drivers. Likewise, vehicle maintenance (which is limited by EPA regulations [*Federal Register 37* (November 15) 1972]) is performed by the manufacturer's experts. There are few limits on the driving conditions during mileage accumulation, so one may assume that the cars are driven in a way that preserves the performance of emissions controls.

Any experienced motorist could formulate a mileage-accumulation procedure more realistic than that now in use. For example, the vehicles might be subjected to a set of drive cycles that would include the current low-speed cruise pattern, but that also would simulate the average teenage driver, months spent idling at stoplights, high-speed freeway commuting, and the shock treatment of many city and country roads. Further, instead of accumulating the mileage in a few weeks, the test fleet would experience a variety of conditions of weather and time, including a minimum of five or six hard winters on salted roads. Some cars would be maintained by the manufacturer's employees, but most would be serviced by a random selection of garage mechanics and service station attendants. Naturally, lack of familiarity with air pollution control maintenance, absence of specialized equipment, or a history of unscrupulous behavior should be no bar to participation in the test.

No doubt, this would be a more accurate test of durability than the one currently in use. Just as surely, this whimsical list makes it clear that it is not possible to implement a test which comes close to the conditions most cars are subjected to over their lifetimes.[5] Prototype vehicles are ready for testing only a bare few months before production begins, and there is not time to do more than accumulate the mileage continuously on a test track. Even if it were possible to simulate time and weather, it is not likely that an officially sanctioned test procedure could simulate the shortcomings of a substantial sector of the service economy.

2. *Failure To Perform Maintenance.* In the absence of some form of state or federal enforcement, it is unlikely that many motorists will adopt the program of periodic emission control maintenance that is implicit in the federal certification test procedure or in the manufacturer's maintenance instructions. This is not to say that cars will not be maintained at all. To the contrary, most cars will get the service and repairs needed to make them run well and insure their resale value. But this type of maintenance and tuning need not contribute to the performance of emissions control devices. To the contrary, with the current catalyst-controlled ICE designs, the natural tendency to tune the engine

---

[5]It has been suggested that the test result for a particular set of prototype vehicles could be adjusted on the basis of surveillance of actual road vehicles from previous model years. The idea is not likely to work in practice. Data on 50,000-mile performance is available only for vehicles that are five or more years old. Given changing vehicle and engine designs, the relevance to new cars of the performance history of vehicles this old would be difficult to establish precisely. The time lag is simply too long to make such an approach feasible.

for best driveability will degrade the performance of the pollution control system. So it is necessary to speak of *emissions control* maintenance when considering the effect of the service system on automotive pollution. No doubt during the preparation for the certification test, this is precisely the type of maintenance a prototype vehicle receives, subject to limits set by the EPA. Cars on the road are not likely to be treated so well.

3. *Tampering.* With projected catalyst-controlled ICE designs, it will be possible to improve both driving characteristics and fuel economy by readjusting the engine timing and carburetion, or by actually disconnecting or otherwise tampering with particular elements of the control system. How strong an incentive there will be to do this is not yet known, for it is not clear how severe the driveablility and fuel penalties will be for vehicles controlled to the 1975-76 standards. Unfortunately, it is easy to imagine rather simple forms of tampering that could boost emissions by a factor of ten to fifteen over the legislated levels for 1975-76 and subsequent years. Tampering will be difficult to control. If it becomes a widespread phenomenon, the effect on the overall air quality control program could be disastrous.

At this point, no one knows how badly emissions control performance will be degraded by the rigors of on-the-road conditions, irregular and unskilled maintenance, and tampering. There have been surveillance studies of on-the-road emissions that indicate there is a real cause for concern [Hocker 1971]. Therefore, for our purposes we need a model that will allow the testing of a number of assumptions about these phenomena. The particular approach we use is shown in schematic form in Figure 6-1, using CO as an example. The bottom line shows the prototype test result $P_v(m)$ for a set of vehicles of a particular model year or "vintage" v; it is the aggregate of a set of emissions functions of the form $e(m)$ in Figure 5-5 for the vehicles submitted for prototype testing. To achieve certification, the average emissions rate must be at or below the standard for $m =$ 50,000 mi (3.4 gm/mi as the figure shows). The other functions represent the mean emissions rates for this group of vehicles under various conditions of maintenance and mileage.

As shown in Figure 6-1, a group of vehicles is considered as being in one of five conditions or "states," $s = 1, \ldots, 5$, depending upon its history of specialized maintenance of emissions control systems. If the vehicles have had emissions control service within the past year, they will be in State 1. The mean CO emissions of the vehicles in that state is denoted $E_{v1}(m)$ as the figure shows. All vehicles start out in State 1, and as long as they receive specialized service at least once per year, they remain in that state as mileage m builds up. The difference between the function $E_{v1}(m)$ and the prototype test result $P_v(m)$ may be due to the "test-to-road" deviation defined earlier. It also may be due to the fact that assembly-line surveillance (discussed in Chapter 5) is not adequate to insure that production vehicles actually match up to the results attained with hand-crafted prototype cars.

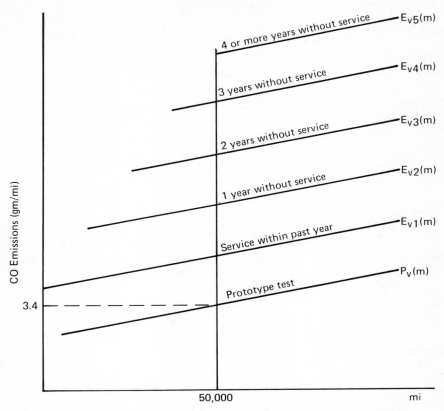

Figure 6-1. Model of Vehicle Emissions as a Function of Miles Driven and Number of Years Without Specialized Service of Emissions Controls

If a set of vehicles of a particular vintage go for one year without emissions control service, then in the subsequent year the group operates at a higher mean emissions rate, as indicated by the function $E_{v2}(m)$. Figure 6-1 is a schematic diagram only; the precise distance between the functions is not of significance. If at the end of the subsequent year the set of vehicles receives maintenance, then it returns to the $E_{v1}(m)$ emissions schedule. As two, three, or four consecutive years pass without specialized maintenance, the mean emissions rate of this particular group moves to the level indicated by $E_{v3}(m)$, $E_{v4}(m)$, and $E_{v5}(m)$. State 5, with emissions rate $E_{v5}(m)$, is the final state a vehicle enters; essentially the assumption is that after four years without service emissions are about as bad as they are going to get. It will be noted in the figure that the functions for States 2 to 5 do not run to the lefthand axis. This is because a vehicle group must be of a certain age before it can be in one of the more deteriorated states, and during these years of aging the average mileage is building up.

Now, using the simple model of Figure 6-1, we introduce the concept of the "stability" of a particular vehicle design. As noted in Chapter 3, a stable vehicle is one that will, on the average, perform near to the prototype test result even without elaborate enforcement systems and special emissions system maintenance. That is, with stable designs the functions in Figure 6-1 are all very close to the lower line $P_v(m)$. An unstable vehicle design is one that experiences significant test-to-road deviation, and that has rapidly increasing emissions as years pass without specialized service. In terms of Figure 6-1, the functions $E_{v1}(m)$ to $E_{v5}(m)$ are widely spread out above $P_v(m)$ in this case.

Naturally, the success of alternative vehicle designs and the cost-effectiveness of mandatory air quality maintenance programs are heavily influenced by the relative stability or instability of vehicles when they get into the hands of the average motorist. In order to study the effects of alternative assumptions about the quality of the control systems on post-1975 ICE-powered vehicles, four stability levels are utilized. These levels are intended to span the range of possible outcomes of current design and manufacturing efforts—from Level 1, which is about as stable as the most optimistic projections, to Level 4, which is about as bad as the forecasts of the most pessimistic observers. We can denote the stability level of a particular group of cars by the index ℓ. Then for any stability level ℓ, the vehicle emissions rate in each maintenance state can be defined as a multiple $D_{\ell s}$ of the prototype test result,

$$E_{vs}(m) = D_{\ell s} \cdot P_v(m). \tag{1}$$

Note that for purposes of this analysis all the vehicles of a particular model year (vintage) are considered to be of the same stability class. In effect it is assumed that the performance of each model year will reflect the then current technology and manufacturing technique, and thus the variations among model types and manufacturers are subsumed in the general group of vehicles of a particular year. Therefore, any vehicle may be classified as being of some vintage (and its associated stability level) and in one of the five maintenance states.

Table 6-1 shows the values of $D_{\ell s}$ used in the analysis reported here and in Chapter 3. The table shows each of the four stability levels relevant to the ICE ($\ell = 1, \ldots, 4$), plus an additional level ($\ell = 0$) which represents a stable alternative technology.[6] A group of ICE-powered vehicles with Level 1 stability starts out with a mean emissions rate only about 20 percent above the prototype test result. Even after four years without service of the emissions control system, the

---

[6]A separate set of values for $D_{\ell s}$ was used for pre-1975 vehicles based on data presented by the National Academy of Sciences [1972].

Table 6-1. Definition of Alternative Stability Levels

Stability Level ℓ	Average On-the-Road Emissions, $D_{\ell s}$, as a Multiple of the Prototype Test Result, for Different Maintenance States, s				
	$s=1$	$s=2$	$s=3$	$s=4$	$s=5$
0	1.1	1.1	1.15	1.15	1.2
1	1.2	1.5	1.75	1.9	2.0
2	1.5	2.5	3.2	3.6	4.0
3	2.0	3.4	4.6	5.4	6.0
4	3.0	5.6	7.4	9.0	10.0

mean emissions rate of the surviving cars is only about twice that predicted by the prototype test. Few observers expect such a fortunate result for the catalyst-controlled ICE. It is essentially a "perfect" system, with only a small deterioration rate and miniscule probability of failure. In addition, vehicle performance is so satisfactory that no one has any incentive to tune engine parameters away from the settings that yield the lowest emissions.

Levels 2 and 3 are significantly worse than 1, with emissions rising to four and six times the prototype performance. This level of degradation of control system performance results from easily conceivable assumptions about catalyst deterioration, perverse maintenance, and tampering with control devices. Level 4 is near the "worst case" for purposes of this analysis. It implies very poor durability of catalysts, and cars that maintain desired emissions parameters only under a continuing regime of emissions-control maintenance. Level 4 also supposes that vehicle performance and driveability are so degraded by the control package that perverse maintenance and tampering take place on a mass scale. In effect, Level 4 implies that the cars controlled to 1975-76 emissions standards may sneak through the prototype test, but on the road, without specialized maintenance, the mean emissions will be only slightly below those from vehicles of 1970-72 vintage.

In effect, the assumptions in Table 6-1 define the vertical distance between the emissions functions in Figure 6-1. If, on the average, vehicles of age a have traveled M_a miles, then the expected emissions rate in calendar year t, E_{vs}^t, can be expressed as

$$E_{vs}^t = [E_{vs}(M_a) + E_{vs}(M_{a-1})]/2, \qquad (2)$$

where $a = t - v + 1$ and $M_0 = 0$.

As emphasized in Chapter 3, no one knows at this point how stable the post-1975 vehicles will be. The stability levels displayed in Table 6-1 are designed to span the range of possible outcomes. Levels 1 and 4 are not likely to come about, but they are possible. Many observers believe emissions performance will be in the neighborhood of Levels 2 or 3. To see how emissions forecasts can be made using this simple model of emission stability, consider the case with no maintenance program. Later the effects of maintenance will be added.

Emissions Forecasts Without Enforcement Programs

In any year, each of the vehicles of any particular vintage will be in one of the five maintenance states s. We denote the fraction of cars of model year v that are in maintenance state s in year t by the symbol S_{vs}^t, and the distribution among states by a column vector S_v^t. It is assumed that all vehicles start out in State 1, that is, $S_v^v = [1, 0, 0, 0, 0]$. As years pass, vehicles move from state to state depending on the nature of the inspection and maintenance system and the characteristics of the cars of a particular model year. We approximate this state-to-state evolution by means of a transition matrix T such that

$$S_v^{t+1} = T \cdot S_v^t. \tag{3}$$

If there is no system of inspection and maintenance, and if there is no incentive for individual motorists to purchase air quality maintenance on their own, then each vintage of cars will simply move from state to state as the years go by. In terms of Equation (3) this means the matrix T is of the form

$$T = \begin{vmatrix} 0 & 0 & 0 & 0 & 0 \\ 1 & 0 & 0 & 0 & 0 \\ 0 & 1 & 0 & 0 & 0 \\ 0 & 0 & 1 & 0 & 0 \\ 0 & 0 & 0 & 1 & 1 \end{vmatrix}, \tag{4}$$

where the columns indicate the states in period t, the rows are the states in period $t + 1$, and the elements of the matrix represent the fraction of cars moving from one state to another. The version of T shown in Equation (4) indicates that all cars simply advance one state each year (until they are trapped in State 5)[7]. For example, all cars that are in State 3 in year t (column 3), will be in State 4 in year $t + 1$, since $T_{43} = 1$.

Equations 1 through 4 form a model of the CO emitted per mile by vehicles of a particular vintage. Next we need to approximate the number of cars

---

[7] Later, when inspection and enforced maintenance are added, the matrix T will be determined by the test instruments and the rules and procedures followed in the inspection station itself, by the emissions standard and stability level associated with a particular vintage of vehicles, and by the effectiveness of the maintenance cars receive if failed in an inspection.

of different model years (vintages) that are going to be on the road in any year, and the mileage they can be expected to accumulate. In the calculations presented here, we work with the national vehicle fleet. Starting with historical data on survival rates and an assumption about the rate of growth of the fleet, we can formulate a model of the age distribution of the car population.[8] The end result of the submodels of the car fleet and its growth is an estimate of the number of vehicles of vintage v on the road in year t, N_v^t. Then a vector expressing the number of cars of vintage v in each of the five states in any year t can be calculated as the multiple of the distribution vector for that vintage S_v^t and the surviving population of that vintage N_v^t—that is, $S_v^t N_v^t$. When multiplied by the average mileage driven by cars of this age—yielding $S_v^t N_v^t M_a$, where $a = t - v + 1$—the result is a vector expressing the number of miles driven in each state by vehicles of vintage v in year t. When these mileage data are multiplied by the appropriate mean emissions *rates* for cars in different states, the result is the total emissions from cars of a particular vintage. Note that E_{vs}^t is the average emissions rate of cars of vintage v in year t if they are in maintenance state s. If we denote the row vector of these emissions rates as E_v^t, then total emissions for a given pollutant (here we still use CO as an example) in year t can be calculated as

$$X^t = \sum_v E_v^t \cdot S_v^t \cdot N_v^t \cdot M_a . \qquad (5)$$

Thus far, we have calculated the gross emissions in a single year of a single pollutant. What is needed for evaluation of alternative policy measures, however, is an aggregated measure of emissions reduction over time, where there is not one pollutant of interest but three. First, we aggregate over pollutants by a weighting function w_k, where $k = 1, 2, 3$ refers to CO, HC, and NO_x respectively. And then we take the undiscounted sum of emissions over a set of years t to yield an aggregate of emissions

$$\bar{X} = \sum_t \sum_k w_k X^{kt} . \qquad (6)$$

In the calculations reported here, a fifteen-year time horizon is used, $t = 1975$, ..., 1989.

The aggregate emissions $\bar{X}$ as calculated in Equation 6 is in physical terms—in "equivalent tons" of some surrogate for auto pollution. For the discussions of policy in this area it is convenient to express these results for any policy as a simple index reflecting the degree of reduction achieved by some policy measure or another. Since the focus in this analysis as a whole is on the *incremental* reduction in emissions associated with the tight standards set for 1975 and 1976, we use the emissions that would be experienced under the 1973 controls as the base

---

[8]The model used here was laid out by T.A. Moen. For a similar model yielding similar results, see Martin [1973].

condition for constructing this index. The degree of cleanup achieved by various measures is called the Weighted Index of Reduction, W. For any policy under study, it is calculated as

$$W = \frac{\bar{X}_{1973 \text{ standards}} - \bar{X}_{\text{policy}}}{\bar{X}_{1973 \text{ standards}}} \qquad (7)$$

One aspect of this model that deserves some elaboration is the choice of the weighting scheme for pollutants w_k. The weighted index W is a measure of "cleanup" and as such must reflect the relative importance of the three pollutants CO, HC, and NO_x. The research on health and other pollution damage discussed in Chapters 7 and 8 is not to the stage where a single set of weights can be determined, and, therefore, we have used three sets of weights in all our calculations.[9] One set, which lies behind all the results in Chapter 3, is related to problems of photochemical smog, and this weighs HC and NO_x more heavily than CO; the weights in this case are w_k = .12, 1.0, 1.0. A second set of weights gives much more significant attention to CO in relation to the precursors of smog by using weights w_k = 0.1, 0.6, 0.3. Finally, we use one set that weights all three pollutants equally, w_k = 1.0, 1.0, 1.0.

The sensitivity of the analysis to assumptions about this weighting function is shown in Table 6-2. The data in the table are for established policy

Table 6-2. Weighted Index of Reduction Under Established Policy Under Alternative Weighting Schemes

Stability Level	Weightings of CO, HC, and NO_x		
	.12, 1.0, 1.0	.1, .6, .3	1.0, 1.0, 1.0
1	.55	.56	.58
2	.45	.46	.48
3	.35	.36	.37
4	.13	.14	.15

(this is the option EST defined in Chapter 3), and results are shown for each of the four possible stability levels, ℓ. As the table indicates, the results are not sensitive to the choice of weighting scheme. In general, the heavier the weight accorded to CO, the higher the index of reduction, but the differences are so small that the choice of one or another weighting scheme, within a reasonable range, does not influence the results of the analysis. This is a fortunate result, for it makes the Weighted Index of Reduction relatively easy to interpret. It is always possible, of course, that some new engine technology or cleanup device may

---

[9]These sets of weights are very similar to those used in studies by the EPA [TRW Systems Group 1971].

have technical characteristics that will make the precise definition of the weighting scheme more important.

Functions of a Maintenance Program

With this emissions forecasting model in hand, we can turn to the issue of state inspection schemes. Our analysis of state enforcement is based on the notion that a system of inspection and enforced maintenance is intended to identify cars that have worked their way into the higher-polluting states and send them back to a cleaner state. In terms of Equations (3) and (4), the function of an enforcement system is to change some of the numbers in the top row of the matrix T to nonzero values, so that some portion of the vehicle population ends up in State 1 each period.

Now since this chapter starts out with the argument that state inspection schemes are likely to be very difficult to implement effectively, and that the maintenance sector may not be up to the task, care needs to be taken to insure that this model analysis gives these systems a fair treatment. Many assumptions must be made along the way in setting up a model of this type—some adding small biases for or against the efficiency of state enforcement schemes. So to be perfectly sure that the analysis casts these systems in a favorable light given the uncertainties, we make an extremely optimistic assumption about what happens when a car fails a state inspection. It is assumed that any group of cars failed in an emissions test and sent for maintenance will emerge from the service garage like new. In terms of the model, any vehicle sent for maintenance, regardless of its age or the state it is in, will be returned to State 1. This assumption is so optimistic about service sector performance as to overwhelm other minor elements of bias in the model.

Naturally, the effectiveness of the enforcement scheme depends also on the efficiency of the inspection program in picking out the high emitters. As discussed in Chapter 5, the short tests that will be used in state emissions inspections do not correlate perfectly with the more expensive (and more accurate) federal certification test procedure.[10] Moreover, there will be errors in the tests themselves, due to unavoidable test variation as well as equipment failure and operator error. This means that whatever the limit of threshold emissions some cars that actually are above the limit will pass the test and others which actually are below the threshold will be failed and sent for repairs they do not need. Unfortunately, at the present time no one knows precisely what kinds of on-the-road emissions tests might be applied to post-1975 vehicles, and so we must build a model that allows flexibility in the choice of assumptions about the possible characteristics of the testing system.

First of all, the test result cannot be too complicated if it is to be

---

[10]The EPA has published a useful summary of the available evidence on this point [1972].

applied to tens of millions or even hundreds of millions of vehicles. There is likely to be some very simple outcome of the instrument readings—perhaps a scalar measurement of some sort based on a simple formula for weighting the results for the different pollutants. Or perhaps there will be a set of two readings, one for some amalgam of CO and HC and one for NO_x. Likewise, the failure level will have to be defined in very simple terms.

Ideally, one would hope that a thorough analysis would be performed in order to determine the failure rate for each vintage and model type. In this way the repair dollar might be spent where it can achieve the greatest reduction in emissions, and the ambient air quality goals would be just met without wasted effort. Unfortunately, all evidence to date indicates that, in order to meet the three criteria outlined earlier in the chapter, the pollution threshold is likely to be set in a much cruder manner. Following the experience in New Jersey [Andreatch, Elston, and Lahey 1971], most states probably will establish the allowable on-the-road emissions rate so that some predetermined percentage of all vehicles will fail the test. No state administration wants to face the prospect of setting an emissions rate and failing two-thirds of all the vehicles tested. The public reaction might do serious damage to the air pollution control program—not to mention the risks to the administrators who are responsible. On the other hand, if the emissions limit is set so low that only a few vehicles fail, then the inspection program is an obvious waste of money—another risky prospect. Moreover, the testing scheme is likely to be set up so that the owners of old and new cars are treated in what appears to be an equitable manner. That is, the test criteria will be set so that the percentage of vehicles failed is roughly the same for all model years.

Our analysis is based on these assumptions about the likely structure of state inspection programs, and calculations have been done for systems that fail 15 percent and 30 percent of all vehicles tested. The results for a 30 percent failure rate are provided in the analysis of Chapter 3. We investigated several models of the life experience of groups of vehicles under such a scheme, considering the test variance problem summarized earlier.[11] The computations shown here are based on a simple model; more complex versions were found to yield substantially the same results. All vehicles are considered to be subjected to an emissions inspection to determine a measured emissions rate e for each car. For each vintage there is a threshold emissions rate $\bar{e}_v$ which divides all the cars into two subgroups—those that pass and those that fail. Naturally, due to test error the true emissions of any vehicle may be above or below the result gained by one single measurement with an inexpensive test. As stated earlier, the cars that fail are considered to be sent for emissions-control service and returned to State 1 from whatever maintenance state they may have been in at the time of inspection. The mean emissions rate for the "failed and serviced" group is calculated by Equation (2) with $s = 1$. The cars that pass the inspection continue to

---

[11] The results of that investigation are summarized in Appleman [1973].

age and deteriorate, and they are considered to move to the next highest state. They operate in that maintenance state for the following year, and emit at a rate expressed by Equation (2), with the *s* value appropriate to the number of years they have gone without service.

The threshold emissions rate for each vintage, $\bar{e}_v$, is calculated (taking into account test variance, initial emissions rates of vehicles, and stability) so that the five-year average failure percentage has a particular assumed value—i.e., 15 or 30 percent as stated earlier. Given these assumptions, it is the case under this model that, for any failure level $\bar{e}_v$, some fraction ϕ_{vs} of the cars in state *s* will pass the test and move on to the next state; the remainder $1 - \phi_{vs}$ will fail. The precise values of the ϕ_{vs} depend on the mean emissions observations for a particular vintage and state, on the test variance (assumed the same for all maintenance states), and on the threshold emissions rate $\bar{e}_v$.

As a result of this process, any group of vehicles of a particular vintage will be distributed over the five states according to the column vector S_v^t in any year *t*. The distribution in the following year will be

$$S_v^{t+1} = T(\bar{e}_v, v) \cdot S_v^t \tag{3a}$$

where

$$T(\bar{e}_v, v) = \begin{vmatrix} 1-\phi_{v1} & 1-\phi_{v2} & 1-\phi_{v3} & 1-\phi_{v4} & 1-\phi_{v5} \\ \phi_{v1} & 0 & 0 & 0 & 0 \\ 0 & \phi_{v2} & 0 & 0 & 0 \\ 0 & 0 & \phi_{v3} & 0 & 0 \\ 0 & 0 & 0 & \phi_{v4} & \phi_{v5} \end{vmatrix} \tag{8}$$

So, for example, if there were no inspection system (and no voluntary maintenance of air quality control devices), $\phi_{vs} = 1$ for all *s* as shown by Equation (4).

We have already suggested how $\bar{e}_v$ will be determined, given the characteristics of the vehicle fleet. Our assumption is that the threshold emissions level will be set so as to fail a certain percentage of the cars of each vintage. If S_{vs}^t represents the elements *s* of the column vector S_v^t in Equation (6), then $S_{v1}^v = 1$ as specified earlier (i.e., all cars start out in State 1), and the variable S_{v1}^t indicates the percent of cars failed for all $t > v$. By iterative calculation it is possible to solve for the value of $\bar{e}_v$ such that the five-year failure rate *R* is at any given level, where

$$R = \frac{1}{5} \sum_t S_{v1}^t, \quad t = v+1, \ldots, v+5. \tag{9}$$

When this analytical model is run for the national vehicle fleet, we obtain the results shown in Table 6-3. Several important aspects of maintenance

Table 6-3. Weighted Index of Reduction Under Alternative State Enforcement Schemes (w_k = .12, 1.0, 1.0).

Stability Level	Percent of Vehicles Failed in State Inspection and Sent for Mandatory Service			
	Full Life of Vehicle			50,000 mi only
	none	15 percent	30 percent	30 percent
1	.55	.57	.59	.57
2	.45	.50	.54	.49
3	.35	.42	.47	.42
4	.13	.25	.34	.25

schemes are evident in the table, and with this information we can elaborate on the discussion of this issue already presented in Chapter 3. Look initially at the first three columns, which present the Weighted Index of Reduction for three cases: no inspection program, a program failing 30 percent of vehicles, and a program failing 15 percent. It can be seen that, at any given stability level, the more cars that undergo enforced maintenance, the greater the reduction in emissions. But the marginal reduction achieved by failing more cars goes down as the failure rate goes up. Raising the failure percentage from 15 percent to 30 percent does not buy as great a reduction as the first 15 percent does, and this pattern continues if the failure percentage is carried above 30 percent. The reason for this behavior is evident: the higher the failure percentage, the more cars are being sent for specialized service that were not bad pollutors to begin with, so more and more of the special maintenance effort is wasted. (By the way, this result argues *against* the schemes, suggested by some observers, that there be no inspection but simply a requirement of emissions control service for 100 percent of cars each year; it is an extremely wasteful approach.)

A second point to notice is that maintenance schemes look more or less attractive depending on how stable the vehicles are expected to be. With Level 1 stability, the gain in the Weighted Index of Reduction is only from .55 to .59 with the institution of an inspection program failing 30 percent of all vehicles. On the other hand, if stability is as bad as Level 4, then the gains are dramatic: the Weighted Index rises from .13 to .34. Even the increase from a 15 percent to a 30 percent failure rate achieves a strong improvement—from .25 to .34. These results once again illustrate the paradox discussed in Chapter 3. If vehicles are stable, then there is little gain from a program of inspection and enforced maintenance. If vehicles are unstable, the maintenance program can have a large positive effect, but the final outcome in terms of pollution still falls far short of the legislated goals. If there were no alternative policy measure (e.g., forcing a basic shift to a more stable automotive technology), then under conditions of poor stability the imposition of these state enforcement programs would clearly be warranted, as argued in Chapter 5. Fortunately, other alternatives *are* available.

Table 6-3 shows one other point about these inspection schemes. All the calculations to date and those reported in Chapter 3 assume that vehicles stay in the inspection and maintenance system for their whole life. A ten-year-old car is treated just like any other at inspection time. But what if cars fall out of the inspection and maintenance network at 50,000 miles or five years of age? The warranty provisions of the act only extend this far, and it is not clear that the Clean Air Act intends that cars should be monitored after this point, or whether such service standards could be enforced under the current law. If it were the case that a motorist could be forced to purchase this specialized service only for the first five years or 50,000 miles of vehicle life, then the benefits of a maintenance program must be adjusted, as shown in the righthand column of the table. If unstable to begin with, these old cars will have a significant impact on overall pollution when dropped out of the enforcement system at the end of the warranty period. How this issue is to be handled has not yet been determined by the EPA.

CONCLUSIONS

This chapter set out to analyze whether or not the EPA and the states should or could implement the in-use inspection and testing provisions of the Clean Air Act. The "could" question was shown to depend upon three factors: financial cost, organizational changes, and public reaction. Each of these elements of state implementation seemed to have significant consequences. But if the benefits of emissions reductions predicate action, then something will have to be attempted, and it's not immediately obvious that in-use testing is any more difficult than other alternatives such as fuel conversion, episodic controls, staggered work hours, or mass transit. The "should" question depends upon the likely benefits to be gained from the incentives for redesign and careful manufacture on the automaker's part and the testing-induced maintenance of the public resulting in aggregate emissions reductions. The incentives were shown to be too far removed from industry decision-makers in terms of personal responsibility, outlook, and time to make the warranty and recall provisions effective forces for proper design or assembly-line quality control. Consequently, it appears that the EPA and the states should not press for in-use testing as long as an alternative technology option is viable. As argued in Chapter 3, it seems much easier and efficient to create and manage the needed technological change rather than a behavioral one.

Chapter Seven

Health Effects of Automotive Air Pollution

William R. Ahern, Jr.

This chapter and the next discuss the value of improvements in health that may result from different levels of reduction in automobile emissions. In this chapter, background information is provided on air pollution, auto emissions, exposure, and the health effects of air pollution in order to highlight the difficulties that are encountered in identifying, separating, measuring, and valuing health damage from auto exhaust. In addition, the analysis underlying current federal policy on auto emissions is described. In Chapter 8 the results of several attempts by others to place dollar values on air pollution reduction are summarized. We then consider a new procedure for estimating the health damage that might be prevented by incremental reductions in auto emissions and describe, step by step, methods for placing a value on resulting health improvements. Chapter 8 concludes with a few remarks about current research on these subjects.

The focus throughout is on the benefits to be gained from incremental reductions in auto emissions in the range from around 75 percent to the legislated 90 percent cutback. If the costs of incremental reductions in this range turn out to be very high, then some estimate of magnitude of the marginal benefits will be useful in considering proposals to postpone or relax current standards. Also, since deterioration of emissions control devices, inadequate maintenance of control systems, disconnection of devices, and other such factors may result in actual on-the-road emissions far above legislated levels, we want to know how serious the health effects of such a poor result would be.

Figure 7-1 is a descriptive model of the effects of auto exhaust. Auto emissions *(AE)* enter the atmosphere, where they are acted on by atmospheric dynamics *(AD)*, such as wind and sunlight, and where they are mixed with other components of the air. Some of these nonautomotive components include emissions from other sources *(OE)*, such as power plants and trucks. The portion of the atmosphere of concern in this analysis is the ambient air *(AA)*, which is de-

140 Clearing the Air

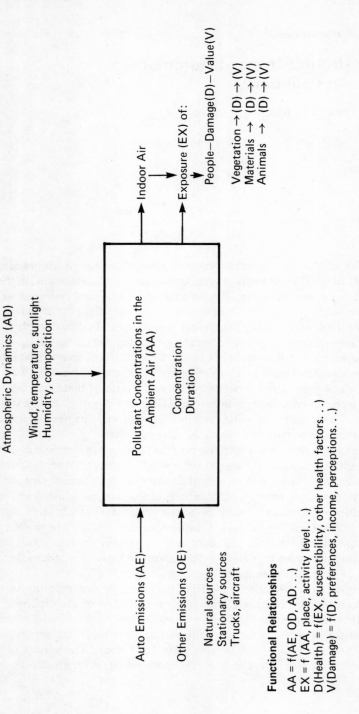

Figure 7-1. Descriptive Model of the Process Relating Health Damage to Auto Emissions

fined as that air to which the general public has access. There is a functional relation

$$AA = f(AE, OE, AD, \ldots),$$

which relates pollutant concentrations in the ambient air to auto emissions, other emissions, atmospheric dynamics, and other factors.

Objects and living things are exposed to concentrations of pollutants in the ambient air. For mobile living organisms the amount of exposure *(EX)* is a function of both location and activity levels, which determine the amount of air they breathe or otherwise come in contact with. Thus, there is another complex relation

$$EX = f(AA, \text{location, activity level}, \ldots).$$

The exposure to pollutants may cause damage *(D)* to the person or object. This damage may be physical or it may be intangible, such as the offending of aesthetic sensibilities. The degree of damage from a particular level of exposure depends on the susceptibility of a person to adverse health effects caused by air pollutants; therefore,

$$D = f(EX, \text{susceptibility}, \ldots).$$

The damage may then be given some value *(V)* by the people affected, by political representatives and government agencies, and by other groups. Such valuation is based on personal preferences and perceptions and on income and other factors. Thus,

$$V(D) = f(D, \text{preferences, perceptions, income}, \ldots).$$

The goals here are to relate valuations of damage to auto emission levels, $V = f(AE)$, and to derive a benefit function for marginal reductions in auto emissions, $\partial V/\partial AE$. As is evident from Figure 7-1, the task is exceedingly complicated.

BACKGROUND ON AIR POLLUTION

Pollutants: Nature, Examples, Interactions

In discussing the complex substances that are products of automobile operation, it is useful to distinguish between an emission and a pollutant. An emission is a substance that is discharged into the air and does not immediately return to earth. A pollutant is a substance that causes some kind of damage or impurity. An emission is not necessarily an air pollutant. For example, autos emit some water vapor and many hydrocarbons that are invisible, odorless, and apparently harmless. Conversely, not all pollutants are made up entirely of emissions. Some air pollutants, especially the photochemical oxidants, are the result of chemical reactions in the atmosphere that transform harmless emissions. Normal atmospheric reactions also eventually convert some harmful pollutants into stable, innocuous substances, such as water and carbon dioxide.

Emissions into the atmosphere come from two broad categories of sources—natural and man-made. Natural emissions result from processes such as pollen production, volcanic eruption, the action of bacterial decay, forest fires, or wind action on surface dust and soil. Man-made emissions result from technological processes, primarily the combustion of fuels. The same type of emissions may come from both kinds of sources. For example, methane gas is produced by vegetation decay in swamps and cattle flatulence, as well as by the combustion of gasoline in motor vehicles.

The two widespread air pollutants that are considered the most toxic are oxides of sulfur and particulate matter. They have been held responsible for the major air pollution disasters, such as the "killer smogs" in Donora, Pennsylvania, in 1948 and in London in 1952. Likewise, the first air pollution emergency declared by the Environmental Protection Agency under the Clean Air Act Amendments of 1970, in Birmingham, Alabama, was due to high particulate levels. The automobile is only a minor source of sulfur oxides and particulate emissions. Nationally, it contributes less than 5 percent by weight of sulfur oxide emissions and less than 10 percent by weight of man-made particulate emissions.[1] Fossil-fuel power plants and industrial smokestacks are the main sources of these pollutants.

Carbon monoxide (CO), a colorless and odorless gas, can also be a toxic pollutant at ambient levels. By weight it comprises about 50 percent of all man-made air pollutant emissions in the country, and over urban areas the auto is the largest contributor of CO emissions [U.S. Department of Health, Education, and Welfare 1970a]. CO is removed from the atmosphere; it does not accumulate in the air. But, to date, the mechanism that accomplishes this removal is poorly understood. In terms of atmospheric reactions, CO is essentially inert. The sink for CO may be atmospheric, biospheric, or oceanic in nature. Its character is the subject of much current research.

The two other principal auto emissions are not pollutants in their emitted states. These are hydrocarbons (HC) and oxides of nitrogen (NO_x). Auto exhaust contains many different hydrocarbons. No harmful effects have been attributed to any of these hydrocarbons at ambient levels. But unsaturated HC molecules, having loose ends (unbonded atoms), are chemically reactive and participate in the formation of photochemical oxidants, which

---

[1]This can vary widely over different urban areas. Autos add about 10 percent of the New York area's particulates, 36 percent of Los Angeles', and 3 percent of St. Louis'. Auto particulate emissions may become an issue of increasing concern as other auto emissions are controlled [U.S. Department of Health, Education, and Welfare 1970c].

are pollutants. The nonreactive hydrocarbons, such as methane, play no role in photochemical reactions in the atmosphere.

NO_x emissions from combustion sources originate in the form of nitric oxide (NO). As with hydrocarbons the NO emissions themselves are not pollutants; no evidence indicates that they have harmful effects at ambient concentrations [Environmental Protection Agency 1971a]. But in the atmosphere, in the presence of sunlight, the NO and the reactive hydrocarbons participate in chemical reactions which form nitrogen dioxide (NO_2) and other photochemical oxidants.[2] Nitrogen dioxide, a brownish gas that is a common visible component of "smog," does affect health. Figure 7-2 shows how NO and NO_2 are involved in a photolytic cycle,

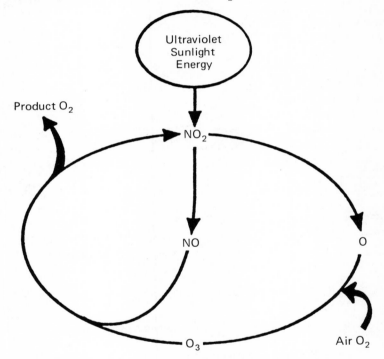

Figure 7-2. Atmospheric NO_2 Photolytic Cycle

Note: Taken from Environmental Protection Agency [1971a]

a chemical reaction in which material is broken down under the influence of light. The cycle, here highly simplified, results in the formation of ozone (O_3), another photochemical oxidant known to be harmful to health.

---

[2] An oxidant is an atmospheric substance that will oxidize certain reagents that are not readily oxidized by oxygen itself. Many are harmful to biological systems and thus are pollutants.

144 Clearing the Air

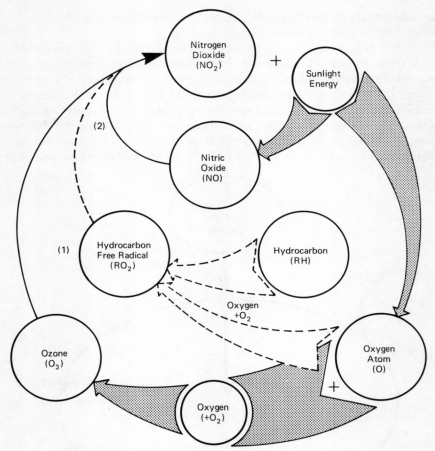

Figure 7-3. Interaction of Hydrocarbons with Atmospheric NO_2 Photolytic Cycle

Note: Taken from Environmental Protection Agency [1971a]

Figure 7-3 shows how increased amounts of O_3 and NO_2 result from the presence of reactive hydrocarbons. Reactive HC molecules are attacked by some of the free oxygen atoms to form very reactive intermediate free radicals (point 1 in the figure). These readily react with O_2 and oxidize NO to NO_2 (point 2 in the figure). Less O_3 is consumed by NO (as there is less NO available) and the NO_2 and O_3 levels increase as the NO decreases.

Other intermediate photochemical reaction products are formed that are pollutants: the eye-irritating aldehydes and the toxic peroxyacetyl nitrate (PAN) are two. The composition of the intermediate photochemical reaction products depends on the concentrations of NO and NO_2 and on the concentrations and reactivity of the types of hydrocarbons involved. Particulates and sulfur oxides (as well as many unknowns) also participate. A primary input is sunlight energy. The greater the light intensity in the wave length range necessary to dissociate NO_2, the more rapid the reaction. Further, oxidation rates of NO and HC increase with any rise in the temperature of the polluted air mass. The rate doubles, for example, with a 40 degree temperature rise.

Formulating a model to relate HC and NO_x concentrations to photochemical oxidant and other product levels is a complicated undertaking. So many variables, such as atmospheric content, dispersion, topography, synergism, and sunlight energy, would have to be included that, given the current state of the art, an accurate model could not be created. Although the major reactions have been duplicated in the laboratory and are fairly well understood, a host of minor but complex reactions also take place, with unknown intermediate products.

Relating Auto Emissions to the Ambient Air

A major difficulty in estimating the damages that can be traced to auto emissions is that a portion of the pollutant concentrations in the ambient air (AA) cannot be isolated and directly attributed to auto emissions (AE). Yet a descriptive model must establish the relationship between auto emissions and air pollution. Factors that make this difficult to accomplish include sparse data measurements, probabilistic variables such as wind that can vary over small areas, and complex interactions among the variables. Furthermore, auto emissions originate from many small sources moving constantly over an urban area. Not surprisingly, the relationship between auto emissions and pollutant concentrations in any place at any time can be treated only in a very simplified way.

Carbon Monoxide. Measurements of CO are reported for eight-hour averages. Aerometric readings of CO vary with the time of day, week, and year, with nearness of sources of CO and with macro- and micrometeorological factors. The highest annual eight-hour readings for major urban areas range from about 12 to 46 milligrams per cubic meter of air (mgm/m^3). The annual median is estimated to be about a third of these values, or 3 to 15 mgm/m^3. The levels in heavy traffic may be higher than these measurements which were taken from air-monitoring stations in the central urban areas, whereas urban residential areas would have lower readings.

Automobiles are the main contributors of CO emissions. Data for 1968 show that motor vehicles account for 59 percent of nationwide CO emissions and for higher percentages in urban areas. Transportation sources, including buses and trucks, contribute about 60 percent of the CO in Philadelphia, about 90 percent in Chicago and Boston, 95 percent in New York, and 99 percent in Los Angeles and the District of Columbia [U.S. Department of Health, Education, and Welfare 1970a].

Nitrogen Oxides. On a global basis, the major source of NO_x is bacterial action. Studies have shown [Environmental Protection Agency 1971a] that in North America the average levels of NO_2 are about 8 micrograms per cubic meter ($\mu gm/m^3$), and the average levels of NO are about 2 $\mu gm/m^3$. But because of the concentration of technological sources, in urban areas the levels are higher. Of the estimated total 1968 nationwide emissions of 20.6 million tons of NO_x, motor vehicles accounted for 7.2 million tons, or about 35 percent. In twenty-two cities, transportation produced an average of 42.6 percent of the total NO_x with the percentage ranging from 23 to 74 percent.

Hydrocarbons. Like NO_x, the major source of atmospheric HC is natural; in this case, organic decay. The background level of unreactive hydrocarbon methane in rural areas ranges from 700 to 1000 $\mu gm/m^3$. Hydrocarbons are reported by their yearly averages of monthly maximum one-hour average concentrations. (A confusing aspect of the study of air pollution is the reporting of measurements for different averaging periods or periods containing a maximum.) Data from the federal Continuous Air Monitoring Project (CAMP) show that over urban areas these HC averages range from 5,250 to 10,130 $\mu gm/m^3$. Unreactive methane accounts for at least half of these amounts [U.S. Department of Health, Education, and Welfare 1970b].

Nationwide technological emissions of HC and related organic compounds for the twelve months of 1968 were estimated to be 32 million tons. Transportation sources accounted for 52 percent of these emissions, with the transportation source percentages ranging from 37 to 99 percent in twenty-two metropolitan areas [U.S. Department of Health, Education, and Welfare 1970b].

Photochemical Oxidants. Photochemical oxidants and the other products of photochemical reactions, since they are intermediate products, do not necessarily appear in urban air near the sources of the emissions that contribute to their formation. The reactions take place in an air mass that usually is moving. Therefore, emissions during the Los Angeles morning rush hour can result in peak oxidant levels an hour later in Azusa and three hours later in Riverside, downwind and to the east. Photochemical reactions are minimal when the sun is not shining and when precursor emissions are low. Therefore, the one-hour O_3 average is at or near zero for about 75 percent of the day. High concentrations in the Los Angeles area occur during less than 10 percent of all hours, and readings can vary widely over an urban area.

Data from twelve monitoring stations for four years indicate that the maximum one-hour oxidant concentrations range from 250 to 1140 $\mu gm/m^3$ with short-term peaks as high as 1310 $\mu gm/m^3$ [U.S. Department of Health, Education, and Welfare 1969a]. Ozone is the main component of elevated oxidant levels, but the composition of oxidants and other reaction products can fluctuate significantly. Attributing portions of photochemical oxidant concentrations to auto emissions, however, is almost impossibly complex since autos are not the only sources of NO_x and reactive hydrocarbons.

Other Pollutants. Although autos are presently considered a minor source of particulate matter, they are the major source of lead in urban air. Gasoline combustion was responsible for about 96 percent of the roughly 187,000 tons of lead emissions in 1968. Food, however, is the principal source of lead intake [Environmental Protection Agency 1971b], and studies have indicated that lead in the air is not a serious health hazard [U.S. Department of Health, Education, and Welfare 1970c]. Further, since lead additives to gasoline probably will be prohibited by federal regulations in the near future, lead is not extensively dealt with here. Other damaging substances, such as asbestos and rubber particles, may be emitted by automobiles, but to date there has been little study of their presence in the ambient air.

EXPOSURE

The second problem in relating auto emissions to health damage, as shown in Figure 7-1, is determining the extent of people's exposure to pollutant concentrations in the ambient air. Exposure has two main dimensions: the concentration of the pollutant and the duration of exposure to that concentration. Measuring the concentration at a number of locations does not give an accurate estimate, since people are quite mobile in a metropolitan area. Furthermore, a person's exposure to air pollutants also varies with his or her rate of breathing; for example, in a heavy activity a person breathes more deeply and faster than usual, thus increasing exposure to pollutants in the air. For an accurate measure of exposure it would be necessary to insert a measuring device in the air passages of every person. Clearly, we shall have to be satisfied with a rough approximation.

Estimating exposure to ambient air is further complicated by the fact that people spend much, if not most, of their time indoors. Indoor air is not ambient air. Air conditioning, space heating, production processes, filters, gas stoves, and chemical no-pest strips may make the character of indoor air different from that of the measured outdoor air. A preliminary study in 1969 [U.S. Department of Health, Education, and Welfare 1970a] showed that household gas-burning devices can cause indoor CO levels higher than those of the outdoor ambient air.

Figure 7-4 presents a simple model of what a commuter's exposure might look like. As he sleeps at night, breathing lightly and with little traffic or industrial activity outside, he is exposed to very low levels of air pollution. Levels become higher as he and others drive to work on congested roads. Indoors, away from traffic, levels drop but may still remain high downtown. A walk to lunch increases his breathing and he is near heavy traffic; therefore, his exposure level rises.

An exposure model that disaggregates the population into individuals is clearly impractical. But the population can be divided into rough geographical exposure categories, such as urban, suburban, and rural populations. Many more air-monitoring stations are needed to provide data for construction of more refined exposure models.

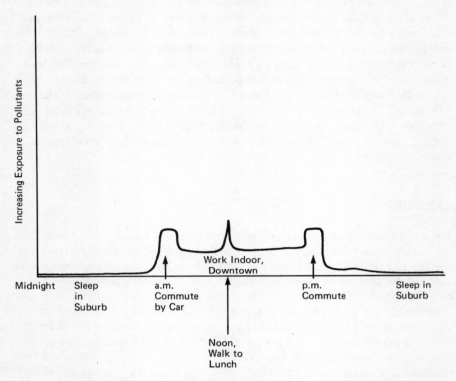

Figure 7-4. Model of a Commuter's Exposure to Air Pollution

BACKGROUND ON HEALTH EFFECTS OF POLLUTANTS

In order to introduce the many complexities involved in assessing health damage from auto air pollution, we will consider the respiratory system, the general short- and long-term health effects of air pollution, the body's adjustments to air pollution, and the role of receptors' health conditions in determining the severity of air pollution's damage to health.

The Respiratory System

The main inpact of air pollution is on or through the respiratory system. The anatomy of the system, diagrammed in Figure 7-5, determines the effects of inhaled gases and particles. Air enters the body through the nasal or the oral cavity. The nasal cavity leads to the pharynx and then to the larynx and trachea, whereas the oral cavity leads directly to the larynx. Thus there is a longer passage for air to travel when it is breathed through the nose.

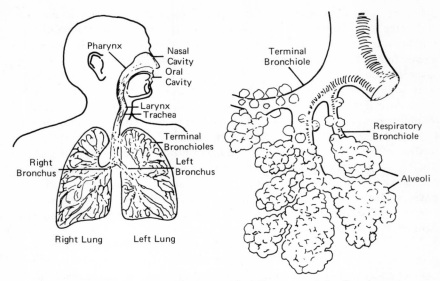

Figure 7-5. The Main Anatomical Features of the Respiratory System (Left) and the Terminal Bronchial and Alveolar Structure of the Lung (Right)
Note: From U.S. Department of Health, Education, and Welfare [1970c]

The trachea leads to the bronchi, a tree-like system of air passages of decreasing size. These bronchi are made up of twenty-three generations of branching tubes. The tubes end in alveolar (air) sacs, about 300 million of them, each about 150 to 400 microns (a millionth of a meter) in diameter. Total alveolar surface, then, may be from 30 to 80 m^2. It is at the alveolar level that the air-blood exchange occurs. Since the cross-sectional area of the system increases with depth, there is a marked decrease in the velocity of air as it approaches the alveoli. The nasopharyngeal and tracheobronchial passages are lined with epithelial cells (a thin layer of closely packed cells that lines the hollow passageways of the respiratory system). There are several kinds of epithelial cells; the ones referred to here are called ciliated epithelial cells because they contain small, hairlike projections capable of rhythmic motion. The flowing movement of these cilia pushes the mucus toward the pharynx where it is swallowed. The final bronchi (called terminal bronchioles) and the alveoli are called the pulmonary structure. Its surface consists of nonciliated, moist epithelium with no protective mucus.

Air pollution may also affect other organs, especially the eyes. The surface of the eyes may be irritated; the skin may also be directly affected. Impacts on other organs, such as the stomach, brain, and kidneys, can only be indirect, resulting from the air-blood transfer or from the swallowing of mucus.

General Short- and Long-Term Health Effects. Air pollutants in general do not act primarily as a *cause* of adverse health effects. Specific air pollutants, such as those from combustion processes, act as irritants to the eyes and to the respiratory system at ambient levels. This irritation can impair bodily functions, especially by reducing the oxygen-carrying capacity of the blood. Such a reduction in the efficiency of the respiratory system can make the body more susceptible to illness. Of course, air pollution is but one of many factors which increase susceptibility to disease; others include inadequate nourishment, fatigue, psychological stress, and the presence of toxic bacteria or viruses, as well as high humidity, extreme temperatures, and rapid temperature change. Several other factors also may exacerbate illness or slow or prevent recovery: the quality of medical care, age, history of illness, and the elements that contribute to susceptibility. Air pollutants primarily exacerbate respiratory illnesses, such as the common cold, sore throat, bronchitis, flu, pneumonia, and pulmonary emphysema. It follows that air pollution may be a contributor to, though not the sole cause of, death, since it may render a terminal illness fatal earlier than under nonpolluted conditions.

The health effects of air pollutants can be separated, roughly, into short- and long-term effects. There is no doubt that high air pollution episodes can cause widespread irritation and contribute to increased illness, exacerbated illnesses, and hastened deaths. Increased mortality and morbidity rates are well documented for the exceptional episodes of air pollution, such as those which have occurred in London, Rotterdam, and New York City [U.S. Department of Health, Education, and Welfare 1969b].

The long-run effects of lower and more common air pollutant concentrations, however, are not clear. First of all, as noted previously, many other factors affect health. Over time, changes in these factors—such as increased standards of living, improved medical care, changing smoking habits, and better sanitation practices—may overshadow the effects of air pollution. To assess the health damage attributable to air pollution, this damage must be isolated from the long-term impact of other important changing conditions. Unfortunately, the effects of air pollutants on health are subtle and difficult to detect, and data about them are sparse and confusing.

The Body's Adjustment to Air Pollution. The body responds to the invasion of pollutants chiefly by trying to rid itself of or inactivate any foreign (or unwanted) matter. For example, when a particle enters the eye, the eyelid responds by blinking and producing tears. These actions are the eye's natural attempt to rid itself of the foreign material. The defense mechanisms of the respiratory system are presented in Figure 7-6. The nasopharyngeal and tracheobronchial structures are protected with mucus. The velocity of the air decreases as it travels toward the lung, and particulate matter is deposited on the mucus. Irritating pollutants may stimulate the production of mucus, which, with the help of the cilia, flows up to the pharynx. There it is swallowed and is subsequently excreted. Pollutant gases may also be absorbed into the bloodstream in the tracheobronchial compartment.

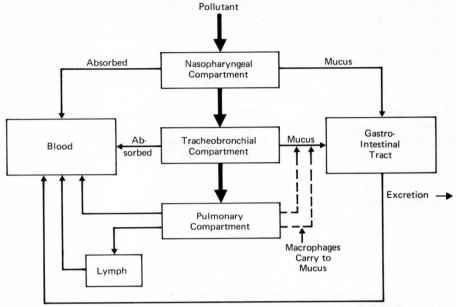

Figure 7-6. Schematic Portrayal of Pollutant Deposition Sites and Clearance Processes
Note: Taken from U.S. Department of Health, Education, and Welfare [1970c]

Clearance in the pulmonary compartment is more complicated since there is no protective mucus. But macrophages, large cells that can ingest microorganisms and other small particles, attack foreign matter and transmit it to the mucus-clearance process. Other pollutants may be removed slowly via the lymphatic system. A large amount of pollutant material is also transferred directly from the pulmonary compartment to the blood, where it can be neutralized.

In addition to defending itself, the body may make adjustments, such as to increase the rate of circulation and raise the blood pressure. These will occur if there is insufficient oxygen in the blood. And, if the bronchi are constricted and the volume of air breathed is therefore reduced, the number of respirations per minute will increase. When any bodily adjustment occurs, however, it is done so at an expense to other organs, such as the heart. The stress on these other organs grows, and they in turn may become less resistant to illness.

Health Conditions of Receptors

In analyzing the health effects of air pollution, it is useful to disaggregate the exposed population on the basis of the receptors' condition. One category consists of receptors who are especially susceptible to the adverse effects of air pollution: the elderly, smokers, and those with respiratory and cardiovascular illnesses, for instance. A general definition of this population might include those persons who need full respiratory efficiency for growth or normal functioning or to prevent exacerbation of illness.

The second category includes all those who are not in the first, that is, the "healthy." These receptors have no special bodily needs, have functional clearance and adjustment mechanisms, and are not very old or very young. The healthy do not require full use of their physiological capacities; therefore, if air pollution slightly impairs their respiratory and circulatory systems, they are not noticeably affected. Of course, in strenuous work or exercise, the performance of the healthy, too, may be impaired by the effects of air pollution. Thus, because of activity or temporary illness, people are constantly crossing between the two receptor categories. And almost everybody is affected by irritation to the eyes and respiratory passages.

There are many ways in which people can improve their health and thereby change their own receptor categories. One way is to increase respiratory and circulatory capacity by exercise, such as jogging or basketball. This raises the level at which air pollutants may have adverse effects. Another way is to avoid debilitating activities, especially smoking. In other words, personal measures may be taken to decrease the severity of the effect of air pollutants.

Assessing the Health Damage Effects of Individual Pollutants

The Clean Air Act of 1967 required the Secretary of Health, Education, and Welfare (HEW)—and in 1971 his successor in the air pollution field, the administrator of the Environmental Protection Agency (EPA)—to provide the states with criteria for setting air pollution standards. The secretary decided to issue a document specifying air quality criteria for each individual pollutant. Industry, academics, conservation groups, and government agencies contributed to the documents. Final drafting and responsibility for content rested with HEW's National Air Pollution Control Administration (NAPCA) and its successor, the EPA. In order of issuance the documents are:

Air Quality Criteria for Sulfur Oxides	January 1969
Air Quality Criteria for Particulate Matter	January 1969
Air Quality Criteria for Photochemical Oxidants	March 1970
Air Quality Criteria for Carbon Monoxide	March 1970
Air Quality Criteria for Hydrocarbons	March 1970
Air Quality Criteria for Nitrogen Oxides	January 1971

Refinements of the documents are planned, but the original versions formed the basis for the national ambient air quality standards set by the Clean Air Act Amendments of 1970.

The first point to stress about the contents of these documents is the difference between toxicologic experiments and epidemiologic studies. Toxicologic experiments are conducted in laboratories where the independent variable—in this case, levels of an air pollutant—can be isolated and where other important factors can be controlled. The experiments are designed to identify the effects of various air pollutant concentrations. Animals are used in experiments involving high dosages: either animals or human volunteers are employed when lower dosages are studied. Many such experiments have guided decisions on air pollution standards for exposure in factories and other places of employment. The toxicologic experiments indicate the physiological mechanisms by which a pollutant, at low levels in ambient air, *may* affect health. In other words, they provide the initial clue for epidemiologic studies of air pollution effects.

The results of some toxicologic experiments are the source of much concern about the serious, long-run effects of certain air pollutants. For example, toxicologic experiments have indicated that repeated exposures to O_3 over a number of months cause lung tumor acceleration in a strain of mice susceptible to such tumors and cause chronic bronchitis in mice, rats, hamsters, and guinea pigs [U.S. Department of Health, Education, and Welfare 1969a]. Such effects have not been discovered in humans at ambient levels of O_3. But such findings naturally generate alarm about possible long-run effects and persuade decision-makers to be conservative in setting standards.[3]

Epidemiologic studies, on the other hand, attempt to find and measure the effects of air pollution on groups of people living in communities, amid all the variables of daily living, where other factors that affect health cannot be controlled precisely, as in a laboratory. The epidemiologic studies take air pollution and other measurements in communities over a period of time and simultaneously try to measure changes in people's health. These changes may be reflected in hospital admissions, cases of flu, school absences, or athletic performance.

The data are then compared, using statistical methods ranging from a simple graph to multiple regression analysis. The results may or may not indicate an *association* of changes in air pollution levels with changes in health indicators. To infer that the air pollution changes *caused* the health changes is a further giant step most epidemiologists are reluctant to take. A major reason for caution is that the pollutant level may be masking another cause, such as a heat wave or differences in smoking habits, occupations, ethnic groups, or methods of indoor heating, or other differences in the populations studied. Health effects may have been caused by such factors as heredity, infections, or allergies. Other pollutants, especially those that are undetected or unmeasured, can also affect the findings. In addition, measurements may be inaccurate or not comparable. And, finally, the researcher may inadvertently omit important factors. For example, in epidemiological studies of the effects of air pollution, smoking is a factor that has often been left out.

---

[3]Similar problems in dealing with alarming toxicologic findings, especially those showing potential cancer or brain damage, beset the officials responsible for setting standards for foods and drugs; witness the cyclamate and hexachlorophene controversies.

An epidemiological study involving admissions of patients with heart attacks to thirty-five Los Angeles County hospitals in 1958 illustrates the problems in drawing conclusions from such studies. These paragraphs appear in the main body of the document on CO criteria:

> Hospital records were abstracted by the medical librarian staff at each hospital. Information obtained included age, sex, date of admission, date of discharge, discharge diagnosis, disposition of patient (recovery or death), area of residence, area of employment, number of days hospitalized, and date of onset of illness. The analysis included 3,080 admissions for myocardial infarction, and involved separate calculations for hospitals in areas of high and low CO pollution.
>
> No significant association was found between the number of admissions for myocardial infarction and ambient CO levels. Significant correlations were found, however, for weekly myocardial infarction case fatality rates and ambient CO levels during the week of admission. Patients admitted to hospitals in the areas of "high" CO pollution where the weekly CO concentration ranged from about 9 to 16 mgm/m^3 (8 to 14 ppm) exhibited statistically significant increases in mortality rates from myocardial infarction when compared to patients admitted during weeks with lower average CO concentrations. This correlation was principally accounted for by an end of the year increase in both case fatality rates and ambient CO levels. To avoid spurious correlations, which can result from day-to-week effects as well as autocorrelation, separate analyses were performed for each day of the week and by high and low pollution areas. Significant associations were then observed only for some of the days of the week and only in the area of the county designated as having high CO levels. The case fatality rates in the high CO area were particularly different from those in the low CO area during weeks of the year with the greatest mean CO levels. The comparability of these two areas was not documented in regard to socioeconomic characteristics.
>
> Factors other than CO exposure, such as hospital admission and hospital-care practices and, most importantly, seasonal influences on case fatality rates, may have accounted for the observed associations. At present, it appears that an association could exist between myocardial infarction case fatality rate and atmospheric CO pollution, but additional studies, particularly of COHb levels in myocardial infarction patients at the time of admission, are required to draw any conclusions about causality. [U.S. Department of Health, Education, and Welfare 1970a, p. 9-13]

Much more research is needed on the health effects of ambient levels of air pollutants in order to permit conclusions about causality. At present the strongest kind of conclusion warranted by such epidemiologic studies is exemplified by this summary of the same study which appears at the end of the CO criteria document:

> There is some epidemiological evidence that suggests an association between increased fatality rates in hospitalized myocardial infarction patients and exposure to weekly average CO concentrations of the order of 9 to 16 mgm/m^3. [U.S. Department of Health, Education, and Welfare 1970a, p. 10-6]

Such language clearly avoids inferring causality. But lay readers unacquainted with the problems and complexities involved may quite naturally read this as, "carbon monoxide at 9 mgm/m^3 kills people with heart problems."

This summary also illustrates another aspect of these criteria documents—successive summarizations distilling a highly complex study into a small paragraph or sentence that gives the impression of certainty about the results. In trying to provide documents with information useful in setting air quality standards, the drafters aimed at compact form and economy of expression. Thus, they looked for the most positive associations between air pollution and possible health damage, and they tended to overlook findings of no effects, uncertainties, qualifications, and inadequacies in the research design.

For example, the first publication of the study of Los Angeles hospital heart attack admissions and CO covered eight pages in the *Archives of Environmental Health*. The main body of the document on CO criteria summarizes this article in the paragraphs just quoted. The final resumé at the end of the document consists of only the one-sentence summary. The listing of a number of such short summarizations at the conclusion of the criteria documents naturally can give the impression that a number of studies have found positive, substantive results associating low air pollution levels with adverse health effects. The documents do, in fact, conclude with statements like this:

> It is reasonable and prudent to conclude that, when promulgating air quality standards, consideration should be given to requirements for margins of safety that would take into account possible effects on health that might occur below the lowest of the above levels. [U.S. Department of Health, Education, and Welfare 1970a, p. 10-6]

Thus, the results of summarization of health effects, combined with the tendency of officials to hedge on the side of safety to protect public health, are highly conservative criteria for setting air quality standards. For example, the cited study on heart attack patients is the only one presented in the CO document that associates a CO level below 12 mgm/m^3 with any adverse effects. Yet the national standard has been set at a strict 10 mgm/m^3 for an eight-hour period.

The following sections outline the background information in the four criteria documents on the major air pollutants associated with the automobile. They summarize the toxicologic findings, the results of two to four key epidemiologic studies of low pollutant levels, and the conclusions.

Carbon Monoxide. Carbon monoxide mixes easily with air and enters the bloodstream where the alveolar sacs exchange gases with the pulmonary capillary blood vessels. When air containing a certain concentration of CO is inhaled for several hours, a state of equilibrium is reached between the CO in the blood and the CO in the air in the lungs. Toxicologic studies show that CO exerts a toxic influence because the oxygen-carrying substance in the blood, hemoglobin (Hb), has an affinity for combining with CO that is 200 times greater than its affinity for oxygen. Carbon monoxide reacts with Hb to form carboxyhemoglobin (COHb) and, when it does, the oxygen-carrying capacity of the blood is reduced.

Since CO causes health damage primarily through the effect of COHb, the toxicologic studies associate physiological responses with the percentage of COHb in the blood. The COHb percentage depends on the concentration and duration of exposure to CO in the air. High COHb levels can result from short exposures to a high concentration of CO or from long exposures to lower levels. For example, human exposure to 35 mgm/m^3 CO for twenty-four hours may result in 5 percent COHb. For long exposures, the increase in COHb is rapid at first but then slows down as an equilibrium is approached. An exposure to 58 to 230 mgm/m^3 for up to two hours may result in 2.5 percent COHb. Some normal processes in the blood result in a background COHb concentration of about .5 percent COHb in healthy persons. Cigarette smokers have a median 5 percent COHb in their blood [U.S. Department of Health, Education, and Welfare 1970a].

The recommendations in the CO criteria document are based on three studies relating low COHb levels to adverse health effects:

1. One study was conducted on eighteen nonsmoking male university students. They were seated in soundproof booths and tested on their ability to compare the length of tone signals as CO concentrations were increased. After ninety minutes of exposure to 58 mgm/m^3 the mean percentage of correct responses fell two standard deviations below the mean attained before CO exposure. Although COHb was not measured, a COHb increase of 2 percent was inferred. Thus, the document summarizes this experiment as follows: ". . . an impairment in timing behavior can be expected to occur at COHb levels of about 2.5 percent . . ." [U.S. Department of Health, Education, and Welfare 1970a, p. 8-18]. But, it notes later that ". . . the subjects were tested in an isolated booth for four hours; boredom or fatigue may well have added to the effect of CO . . ." [U.S. Department of Health, Education, and Welfare 1970a, p. 8-20].

2. Four subjects were tested for visual ability at various levels of brightness. Placed in isolated booths for two-and-one-half-hour sessions, they were exposed to increasing levels of CO. A 4.4 percent degree of impairment was noticed in their ability to perceive relative brightness after a fifty-minute exposure to 58 mgm/m^3 CO. This amount of exposure was estimated to result in a 3 percent increase in COHb [U.S. Department of Health, Education, and Welfare 1970a].

3. The third study was an experiment on forty-nine healthy male adults aged twenty-five to fifty-five who were exposed to about 115 mgm/m^3

for periods of time sufficient to produce COHb levels up to 20.4 percent. A set of sixteen psychological and physiological tests was used to evaluate effects. No control was made for smokers, who made up the majority of the group. The number of errors on half the tests appeared to increase with the COHb level. The experimenter judged that an effect should be detectable at 2 to 3 percent COHb, but the measuring method could not discriminate COHb levels below 5 percent [U.S. Department of Health, Education, and Welfare 1970a].

Many other toxicologic studies cited in the CO criteria document show no adverse effects at COHb levels well above 10 percent, and two subjects with 50 percent COHb ". . . were able to concentrate on their tests and do reasonably well" [U.S. Department of Health, Education, and Welfare 1970a, p. 8-20]. At this point it is appropriate to repeat that the results of these toxicologic studies depend on the testing method and conditions used, and that extending their conclusions to low CO levels and the general population must await epidemiologic confirmation.

Epidemiologic studies of the direct health effects of atmospheric CO have been rare. Many studies of ambient CO, however, do try to link it with increases in COHb. For example, a five-hour exposure of traffic policemen in Paris to 12 to 14 mgm/m^3 of CO increased COHb levels in nonsmokers by about .7 percent [U.S. Department of Health, Education, and Welfare 1970a]. The effect of smokers may be in either direction: smokers with COHb levels above 5 percent may experience a decrease when exposed to atmospheric CO [U.S. Department of Health, Education, and Welfare 1970a]. The study on myocardial infarction (heart attack) admissions to Los Angeles County hospitals in 1958 is cited in the summary as associating low CO levels with adverse health effects. Another study looked at blood samples from 1,518 persons thought to be responsible for auto accidents and compared those blood samples with others from people occupationally or domestically exposed to CO. The possible accident-responsible population had slightly higher COHb levels, but no control was made for alcohol or smoking, so no firm conclusions can be drawn from the study [U.S. Department of Health, Education, and Welfare 1970a].

Cigarette smoking results in comparatively high exposure to CO. When a cigarette is smoked, average concentrations of CO inhaled are 460 to 575 mgm/m^3, and the smoke itself contains more than 22,400 mgm/m^3 of CO [U.S. Department of Health, Education, and Welfare 1970a]. These levels are higher than CO levels encountered in the ambient air. Thus, impairment of bodily functions attributed to CO air pollution could not exceed impairment caused by cigarette smoking.

Studies have shown that compensatory mechanisms develop after long exposure to low levels of CO, much as people adapt to living at high altitudes. Increases in production of Hb, in blood volume, and in the number of red blood cells have been observed in experiments on animals when they are exposed to low levels of CO for extended periods of time [U.S. Department of Health, Education, and Welfare 1970a].

The judgment on COHb is that just as it increases with CO exposure, it also decreases back to body levels when exposure declines. Thus, the effects of CO seem to be primarily short-term.

Hydrocarbons. Hydrocarbons and oxides of nitrogen only affect people's health indirectly, through their activity in photochemical reactions. The photochemical oxidants and other photochemical products of these reactions are the substances that cause health damage. The HC document states:

> Studies conducted thus far of the effects of ambient air concentrations of gaseous hydrocarbons have not demonstrated direct adverse effects from this class of pollution on human health. [U.S. Department of Health, Education, and Welfare 1970b, p. 8-5]

Aldehydes. The HC document contains a section on the health effects of a class of photochemical products called aldehydes, which contribute to the eye irritation caused by photochemical smog. There is a direct relation between aldehydes and HC through various atmospheric reactions. The two aldehydes studied in the toxicologic experiments mentioned in the document are formaldehyde and acrolein. Each can irritate the eyes, upper respiratory system, and skin. Exposure of humans to increasing levels well above those found in ambient air have resulted progressively in lacrimation, coughing, sneezing, headache, weakness, bronchitis, and other respiratory impairments. Experiments on animals indicate that aldehydes have possible synergistic effects with aerosols that make them more toxic [U.S. Department of Health, Education, and Welfare 1970b]. No epidemiologic studies have been done since it is difficult, if not impossible, to isolate the effects of aldehydes from those of other photochemical products.

Because irradiated auto exhaust from a variety of gasoline mixtures is composed partly of aldehydes, it is discussed in the aldehyde section of the HC document. Some data indicate that combustion of fuels high in aromatics and olefins results in reaction products with higher potential for eye irritation than combustion of other types of fuels [U.S. Department of Health, Education, and Welfare 1970b]. The national ambient air quality standard for hydrocarbons is based on their role in producing photochemical oxidents. The standard is 160 $\mu gm/m^3$, a maximum 6:00 to 9:00 a.m. concentration not to be exceeded more than once a year [*Federal Register 36* (April 30) 1971].

Nitrogen Oxides. The NO_x document states, "no evidence shows that NO produces significant adverse health effects at the ambient atmospheric concentrations thus far measured" [Environmental Protection Agency 1971a, p. 11-10]. Nitric oxide (NO) is the auto emission of the oxides of nitrogen, and it does not seem to be a pollutant; but it can be oxidized to become nitrogen dioxide (NO_2), a photochemical oxidant. The nitrogen oxide document deals with the health effects of NO_2. Toxicologic experiments show that NO_2 exerts its primary effect on the lungs. Most animals die from pulmonary edema when exposed to very high concentrations of NO_2 [Environmental Protection Agency 1971a]. Lung tissue damage and pathological lesions have resulted from short-term exposure to high dosages. Exposure to nonlethal doses causes faster breathing, increased airway resistance, and decreased breath volume. A possible explanation for these phenomena is that the acidic environment produced by exposure to NO_2 may inhibit the metabolic activity of the lung tissue. Most of the reactions seem to be reversible when the animals are returned to clean air.

Exposure to NO_2 also causes increased susceptibility to bacterial pneumonia and influenza infections. The test animals show a reduced ability to clear infectious agents from their lungs. Exposure to long-term concentrations below those producing acute inflammation results in a cumulative effect: lesions appear, the deep lung tissue becomes inflamed, and fewer cilia are present. A few occupational studies have been done on humans, and experimental exposure to 9,400 $\mu gm/m^3$ for ten minutes produces a large but temporary increase in airway resistance. Very high accidental exposures to NO_2 fumes have led to pulmonary edema and death in humans.

Epidemiologic appraisal of NO_2 is nearly nonexistent. It relies on two Chattanooga studies that relate ambient NO_2 to respiratory function and illness. The first of the studies compares four different areas in the city of Chattanooga, Tennessee. One area with a high NO_2 level is close to a large TNT plant, which is a source of NO_2 and nitrates. Another area has high particulate levels, and two other areas served as "clean" controls. A total of 987 second graders were tested for ventilatory function in November 1968 and March 1969. Socioeconomic factors were controlled. The study concludes that ". . . the ventilatory performance of the second-grade school children in high NO_2 exposure areas was significantly lower than performance of children in the control areas" [Environmental Protection Agency 1971a, p. 10-3].

A later phase of the same study questioned the families of the children, a total of 4,043 individuals, on frequency of colds, sore throats, and other respiratory illnesses during that winter. The illness rates for the families in the high NO_2 area were consistently and significantly higher than the rates in the two control areas. The average NO_2 concentration in the high NO_2 area, when increased occurrence of acute respiratory disease was observed, ranged from 117 to 205 $\mu gm/m^3$.

The second Chattanooga study was a retrospective investigation covering three of the four areas used in the first study. Subjects were infants and first and second graders. Their parents filled out questionnaires on respiratory illnesses contracted by the children from July 1966 through June 1969. The findings showed a significant increase in acute bronchitis in the high and intermediate NO_2 areas. The mean NO_2 concentration ranged from 118 to 156 $\mu gm/m^3$ in the two high areas. Incidence of croup and pneumonia and hospitalizations for acute lower respiratory illness, however, did not vary with the area.

From these studies the NO_2 document concludes that "any site that exhibits a concentration of 133 $\mu gm/m^3$ or greater exceeds the Chattanooga health-effect-related NO_2 value" [Environmental Protection Agency 1971a, p. 10-8]. The national primary and secondary air quality standard for NO_2 has been set at an annual arithmetic mean of 100 $\mu gm/m^3$ [*Federal Register 36* (April 30) 1971].

Photochemical Oxidants. The document on photochemical oxidants [U.S. Department of Health, Education, and Welfare 1969a] deals mainly with two substances, O_3 and oxidants. Toxicologic studies on animals show that O_3 exerts a toxic effect primarily on the respiratory system. But the mechanisms that produce this effect remain unidentified. High concentrations of the gas may make breathing more difficult, produce pathological changes such as bronchitis and emphysema, chemically change the structure of lung proteins, and increase susceptibility to infections.

Some experiments with mice indicate that the body may develop a tolerance to O_3 and that this tolerance may even protect the respiratory system from other pollutants. Deep lung irritants (such as O_3) can trigger tolerance to other pollutants, whereas more soluble gaseous irritants that do not reach the deep lung (such as sulfur dioxide) cannot. The studies on mice have shown, for example, that exposure to O_3 can protect the respiratory system from subsequent adverse effects of exposure to NO_2. Such tolerance, however, does not develop in baby chicks, and it is not known whether or not humans develop this protection.

Occupational and experimental studies have been done on the effects of O_3 on humans. Concentrations up to 780 $\mu gm/m^3$ for an hour have not produced responses. Exposure to an O_3 range of 980 to 1,960 $\mu gm/m^3$ for one to two hours produces changes in pulmonary function, such as increased airway resistance. Concentrations above 1,960 and up to 5,900 $\mu gm/m^3$ produce extreme fatigue and lack of coordination and become intolerable to some people. Finally, concentrations of about 17,600 $\mu gm/m^3$ cause severe pulmonary edema.

On the other hand, prolonged exposure to lower concentrations (up to 390 $\mu gm/m^3$) results in no observed effects. The threshold level for nose and throat irritation seems to be about 590 $\mu gm/m^3$. And 980 $\mu gm/m^3$ for three hours a day, six days a week, for eight weeks causes a 20 percent decrease in forced expiratory volume. Subjects return to normal in six weeks after the cessation of exposure.

Mixtures of oxidants for toxicologic studies have been obtained either from the air of Los Angeles or other cities or from irradiated auto exhaust. In addition to O_3 these mixtures contain CO, oxides of nitrogen, and HC, so that measured effects cannot be attributed solely to oxidants. Thus, although increases in lung flow resistance were observed in guinea pigs exposed to mixtures of oxidants in Los Angeles (as oxidant levels reached 980 $\mu gm/m^3$), increases also occurred on days when oxidant levels were considerably lower and weather was cold and wet.

Other experiments have attempted to determine whether there is a link between oxidants and cancer or early death. The combination of ozonized gasoline and influenza virus has produced carcinomas in mice. "No true lung cancers have been reported, however, from experimental exposures to either ozone alone or any other combination or ingredient of photochemical oxidants" [U.S. Department of Health, Education, and Welfare 1969a, p. 8-33]. Another experiment indicates that high levels of irradiated auto exhaust, with oxidant levels of 590 to 1,960 $\mu gm/m^3$, result in increased death rates for neonatal mice. The authors of the experiment suggest that ". . . air pollutants in irradiated auto exhaust may alter the genetic composition and possibly other cellular components of sperm" [U.S. Department of Health, Education, and Welfare 1969a, p. 8-33].

Toxicologic studies of the effects of oxidants on humans have focused on eye irritation, the obvious reaction to high ambient levels of photochemical oxidants. Ozone, the principal oxidant in the air, is not an eye irritant, however. Except for aldehydes, the substances in the air that cause eye irritation have not been clearly identified, and no physiological mechanisms indicating a direct cause-and-effect relationship have been demonstrated.

Epidemiologic studies of the effects of oxidants are exceptionally difficult to analyze. One reason is that results are erratic because oxidants reach their peak in the late morning or early afternoon after the sun has acted on the emissions from the morning rush hour. Also, the combination of substances in photochemical oxidants may vary widely, and different oxidants have different effects. Weather conditions affect both oxidant formation and health. Thus, findings attributing health damage to oxidants may be tenuous at best. Studies of various areas in California have attempted to link either mortality or morbidity rates with oxidant levels.

The criteria document on oxidants relies on three epidemiologic studies in its concluding section: one on impairment of athletic performance, one on aggravation of asthma, and one on eye irritations [U.S. Department of Health, Education, and Welfare 1969a]. The first of these, on the impairment of athletic performance in twenty-one cross-country meets by San Marino, Los Angeles County, high school runners from 1959 to 1964, found a statistically significant relationship between the percentage of team members whose performance declined from the previous meet and the oxidant levels one hour before the meets. These oxidant levels ranged from 60 to 590 $\mu gm/m^3$. But a lower correlation was found using oxidant levels for the hour of the race [U.S. Department of Health, Education, and Welfare 1969a].

A second study deals with the correlation of aggravated asthma conditions in the presence of oxidants. Between September 3 and December 9, 1956, 137 asthmatics in Pasadena, all long-time residents of the Los Angeles area, kept records of their attacks. Peak time for attacks was between midnight and 6:00 a.m., when oxidant levels are usually zero. The statistical relation between the number of attacks and the peak oxidant level per day was poor (correlation coefficient of 0.37). But there was a statistically significant increase in the number of attacks when maximum oxidant levels during the day exceeded 250 $\mu gm/m^3$. This relation appeared most strongly with people who had lived in the area for more than ten years. Eight of the 137 asthmatics, or 6 percent of the total, were identified as having the attacks strongly correlated with the 200 $\mu gm/m^3$ level and above [U.S. Department of Health, Education, and Welfare 1969a].

The third epidemiologic study in the oxidants criteria document is a study of eye irritation caused by oxidants in two groups (each consisting of twenty female telephone company employees) in two adjacent rooms in downtown Los Angeles for 123 work days from May to November of 1956. Active and dummy air filters were switched periodically between the two rooms, and the sensory responses of the subjects were measured each morning. Oxidants and other pollutants in each room were also measured. The eye irritation was measured on a subjective index from "none" to "barely noticeable" through "severe." The differences in irritation between the filtered and nonfiltered rooms were highly significant. And the results indicated that " . . . eye irritation for the study groups increased progressively as oxidant concentrations exceeded 200 $\mu gm/m^3$ [U.S. Department of Health, Education, and Welfare 1969a, p. 9-18].

Another dimension of photochemical oxidant effects is that peak oxidant levels, the daily maximum readings, are associated with a lower hourly average concentration. These hourly averages are taken from observational data. Thus, in all three studies the peak value of 250 $\mu gm/m^3$, which occurs for only a few minutes in the ambient air, is associated with a lower hourly average concentration of from 100 to 120 $\mu gm/m^3$. The national ambient air quality standard for photochemical oxidants, taking into account a margin of safety, has been set at 160 $\mu gm/m^3$, maximum one-hour concentration not to be exceeded more than once per year [*Federal Register 36* (April 30) 1971].

The preceding summary of research described in the air quality criteria documents illustrates how difficult it is both to identify and to measure the health damage that may be caused by air pollution. Little research has been done on the effects of combinations of pollutants as they exist in the ambient air. Studies are difficult to conduct because the effects on individuals may be small; because air pollution data are scarce and cover only limited areas of cities; and because other factors, such as nutrition and medical care, have a much greater impact on people's health. Again, estimates of health damage can be made only at a generalized and simplified level, given the current state of knowledge. It is another jump in complexity, then, to separate out from the health damage attributed to air pollution that part which may be attributed to auto emissions. Few studies have attempted this approach.

THE ANALYSIS BEHIND CURRENT EMISSIONS STANDARDS

One of the strongest guiding forces behind the air quality criteria documents and the national air quality standards has been the work of Dr. Delbert Barth, former Director of the Bureau of Criteria and Standards of the National Air Pollution Control Administration of the Department of Health, Education, and Welfare, and presently Director of the EPA National Environmental Research Center in Las Vegas, Nevada. His article, "Federal Motor Vehicle Emission Goals for CO, HC, and NO_x Based on Desired Air Quality Levels" [Barth et al. 1970] had a significant influence on the 1970 Clean Air Act Amendments, which require a 90 percent reduction of CO, HC, and NO_x in auto emissions.[4] The model selected by Barth and his colleagues was a simple "rollback" equation. Their objective was to find out to what extent 1980 auto emissions would have to be controlled in order to reach a specified level of desired air quality (DAQ) in 1990. Vehicles of the 1980 model year and beyond would have to meet the 1990 control requirements, since the car population turns over only about every ten years.

To get the desired percentage reduction in auto emissions, the federal scientists first selected the highest ambient reading in any city for 1967 for each pollutant that is attributed to autos. This maximum reading, regarded as present air quality (PAQ), was then multiplied by an "emissions growth factor" to determine what the PAQ would be in 1990 with no reduction in auto emissions. The emissions growth factor was assumed to be directly proportional to the expected growth in the car population between 1967 and 1990. An EPA internal study [Kramer and Cernansky 1970] calculated this factor to be 2.18.

The DAQ for each pollutant was derived from the air quality criteria documents. The goal of the analysis, as required by legislation, was to provide standards for the maximum possible protection of health. Therefore, in the Barth study the DAQ was chosen at a level below which no study had previously indicated the possibility of adverse health effects.[5] This DAQ was then subtracted from the projected 1990 air quality (1967 maximum level times 2.18) to get the amount of desired reduction in ambient air pollutant concentrations. This amount was divided by the 1990 projected maximum pollution level to get the fraction, or percentage, of reduction required. The Barth study subtracted an atmospheric background level of the pollutant from this denominator. In summary, the equation for calculating the desired percentage reduction in maximum ambient levels of each air pollutant is:

$$\text{Percentage Reduction} = \frac{(2.18 \times \text{max 1967 reading}) - \text{desired air quality}}{(2.18 \times \text{max 1967 reading}) - \text{background level}}$$

---

[4] A letter from the Automobile Manufacturers Association (AMA) to HEW Secretary Elliot Richardson, August 27, 1970, cited this article as the basis of the 90 percent figure. The AMA had received a copy of the article from a member of Senator Muskie's staff. The letter is reproduced in the Air and Water Pollution Subcommittee hearings for 1970 [U.S. Senate, Committee on Public Works, Subcommittee on Air and Water Pollution 1970].

[5] Barth's article predates the National Air Quality standards, but the numbers are similar. See the last section in this chapter for a comparison.

164 Clearing the Air

Then, assuming that the maximim pollutant reading is wholly attributable to auto emissions, the emissions rollback required to reach DAQ in 1990 was calculated. Subtracting the required percentage reductions in emissions from 100 percent gave the percentage of 1967 emissions that could be permitted in 1990. The desired emission rate, in grams per vehicle mile (gm/mi), for cars on the road in 1990 was then the permitted percentage times the 1967 emission rate for each pollutant:

Desired Emission Rate = (100 percent − percent reduction) x 1967 rate.

The Analysis of Carbon Monoxide

Application of these two equations to CO is straightforward. For DAQ Barth selected 10 mgm/m^3 average eight-hour ambient concentration of CO (the current national standard). This value is below the 12 to 17 mgm/m^3 range cited in one experiment as producing an increase in blood carboxyhemoglobin of 2 to 2.5 percent. It is within the 9 to 16 mgm/m^3 range cited as contributing to increased deaths from heart attacks among patients in Los Angeles [U.S. Department of Health, Educatoin, and Welfare 1970a].

The maximum 1967 reading (PAQ) was Chicago's highest eight-hour average of 51 mgm/m^3 and the background concentration was 1 mgm/m^3. The average auto emissions level of CO used was 82.6 grams per mile.[6] The calculations can be written as

$$\text{Percentage Reduction} = \frac{(2.18 \times 51) - 10}{(2.18 \times 51) - 1} = 92.5\%$$

Allowable Emissions Percentage = 100 − 92.5 = 7.5%

Desired Emission Rate = 7.5% x 82.5 gm/mi

= 6.16 gm/mi allowable emission to reach DAQ.

Thus, a 92.5 percent reduction from 1967 auto emissions of CO would be needed in 1980 to insure that the level of 10 mgm/m^3 would not be exceeded in any city in 1990. A descriptive model of this analysis is shown in Figure 7-7. The Barth study, as the figure shows at point 1, assumes that all of the highest CO reading is attributable to auto emissions. This highest reading represents PAQ (shown as point 2) and DAQ is chosen at a level (3) that assures no health damage. The difference between 2 and 3 is the desired reduction in ambient air CO concentrations (shown below the graph as 4). This particular descriptive model is useful for examining the assumptions and simplifications in this analysis.

---

[6]These emissions rates were taken from an internal EPA document [Kramer and Cernansky 1970]. The emissions test used is unknown.

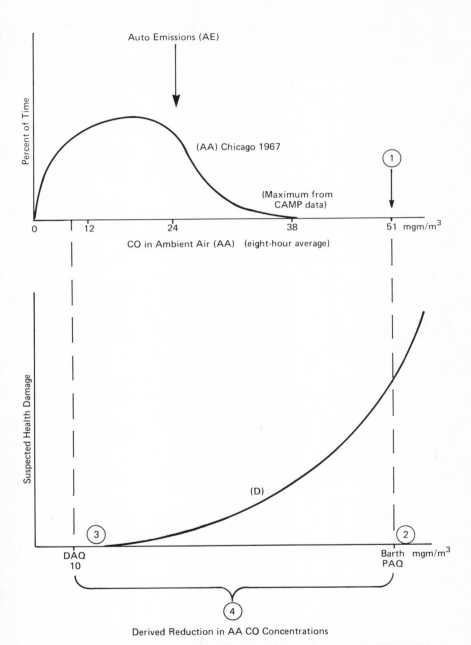

Figure 7-7. Descriptive Model of the Barth Analysis, Taking CO as an Example (CO Distribution from CAMP Data)

First, the model assumes that all the ambient air concentrations of CO are contributed by auto emissions. This eliminates the complex problem of atmospheric modeling and of assigning a part of ambient pollutant concentration to auto emissions, but it overstates the auto's role in CO pollution. In Chicago, all forms of transportation, including trucks and buses, not just autos, account for about 90 percent of CO emissions by weight [U.S. Department of Health, Education, and Welfare 1970a]. Autos would therefore have contributed about 80 percent or less of the CO in the highest Chicago reading.

Second, the measure of exposure is taken as the worst eight-hour average reading in 1967 for any city. This was 51 mgm/m^3 or 44 parts per million (ppm). The CAMP data summarized from Chicago in Table 7-1 indicate that 44 ppm is an unusually high reading, which may reflect unusual circumstances or measurement error. It was chosen, however, because using a maximum pollutant reading eliminates the complex task of modeling exposure of the population to various concentrations and durations of pollutants in the ambient air. It also hedges very strongly on the side of protecting the public health.

Another margin of safety is incorporated in the calculation of the emissions growth factor of 2.18. The method chosen to derive this number assumes that a 2.18 increase in total emissions tonnage resulting from 2.18 times as many cars on the road in 1990 as in 1967 would result in a 2.18 increase in the highest ambient air pollutant concentration. The highest 1967 readings used by Barth, however, are for downtown areas where traffic saturation exists or is being approached. And the 2.18 increase in the number of autos is reflected not only in increases of traffic density but also in the sperading out of traffic and emissions over an expanding metropolitan area. Thus, the assumption of a linear relation between auto population growth and maximum pollutant concentration results in an overestimation of the expected 1990 maximum pollutant concentration. In every case, then, the officials charged with protecting the public from air pollution have understandably made simplifications and assumptions that incorporate large margins of safety to assure that CO will cause no adverse health effects.

Analysis of Hydrocarbons and Oxides of Nitrogen

Carbon monoxide presents few complex problems in relating DAQ to health because it acts directly on the body. Hydrocarbons and oxides of nitrogen emissions, however, cause health damage primarily through their role in formation of photochemical oxidants. The federal analysis uses data reported in the air quality criteria documents to relate HC and NO$_x$ with photochemical oxidants.

This analysis of air pollution measurements shows that—primarily because of atmospheric dynamics—a particular oxidant level can result from a wide range of HC and NO$_x$ concentrations. But the maximum attainable

Table 7-1. Carbon Monoxide Concentration (in ppm) at CAMP Sites for an Eight-Hour Averaging Period, 1962-1967

City	Maximum Reading	Percent of Time Concentration Is Exceeded							
		0.001	0.01	0.1	1	10	30	50	70	90
Chicago	39			35	27	21	16	12	8	5
Cincinnati	18			18	14	8	6	5	4	2
Denver	37			29	18	12	9	7	6	3
Los Angeles	32			27	21	15	12	10	9	8
Philadelphia	31			27	19	12	8	7	5	3
St. Louis	21			17	14	10	7	6	4	3
San Francisco	17			13	11	8	6	5	4	3
Washington, D.C.	26			22	16	8	5	4	3	2

Note: From U.S. Department of Health, Education, and Welfare [1970a]

photochemical oxidant level is limited by the available HC and NO_x precursors, regardless of how stable the atmosphere is. The federal analysis chose this upper limit, the analog to the maximum CO reading, in order to protect public health. All monitoring data were analyzed to relate HC and NO_x individually to maximum daily one-hour average oxidant concentrations. The precursor readings were averaged for the hours of the morning rush hour, 6:00 to 9:00 a.m. Maximum oxidant levels occurred about two to four hours later, after photochemical reactions had occurred in the atmosphere.

HC and Oxidants. Figure 7-8 relates nonmethane hydrocarbons to maximum daily one-hour average oxidants. Data were taken from Continuous Air Monitoring Project (CAMP) stations, one each in downtown Philadelphia, Denver, and Washington, D.C., and from the Los Angeles network, during times of maximum possible oxidant levels (May through October) for the years 1966 to 1968. The upper limit of oxidants was recorded on only one percent of the 125 days of data readings so it is defined by a small number of points on the graph [Environmental Protection Agency 1971a]. A curve was used to predict the maximum one-hour average oxidant concentration from a measured 6:00 to 9:00 a.m. average HC concentration. Thus, a concentration of 200 $\mu gm/m^3$ of oxidants was associated with 200 $\mu gm/m^3$ nonmethane hydrocarbons. Oxidant levels below 140 $\mu gm/m^3$ and HC levels below 200 $\mu gm/m^3$ were not reported because they are subject to measurement errors [Environmental Protection Agency 1971a].

NO_x and Oxidants. The same method was used to predict maximum daily one-hour average oxidants from average 6:00 to 9:00 a.m. NO_x measurements. Data were taken from downtown Philadelphia, Denver, and Washington, D.C., CAMP stations for roughly the same period, but they were not taken from Los Angeles. (Figure 7-9 graphs the data.) Again, one percent of the oxidant points define the upper limit. The data indicate that to keep oxidants below 200 $\mu gm/m^3$ on 99 percent of the days, 6:00 to 9:00 a.m. average NO_x levels must be kept below 80 $\mu gm/m^3$. There are many analytical and measurement uncertainties in this range; the number of points defining the upper limit is insufficient for statistical analysis [Environmental Protection Agency 1971a].

HC, NO_x, and Oxidant Levels. As mentioned in the section on photochemical reactions, reactive hydrocarbons and NO_x act together to produce photochemical oxidants. The same data base of 125 days of observations for one station each in the three cities was used to graph (in Figure 7-10) the combinations of hydrocarbons and NO_x, 6:00 to 9:00 a.m. averages, with identical one-hour maximum daily oxidant concentrations. Only the upper limit points were used. Each curved line represents one maximum oxidant level, with levels increasing to the upper right. This indicates that the maximum possible oxidant level depends not only

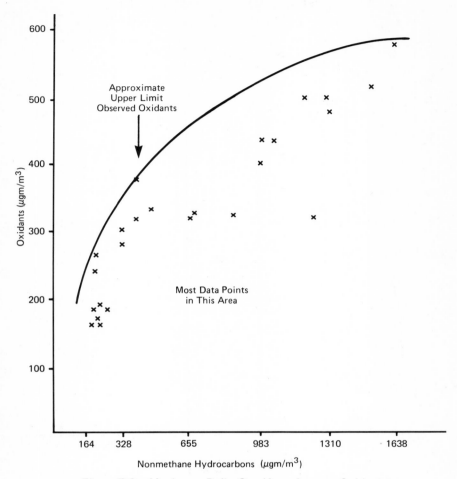

Figure 7-8. Maximum Daily One-Hour Average Oxidants as a Function of 6:00 to 9:00 a.m. Averages of Nonmethane Hydrocarbons at CAMP Stations, June through September, 1966 through 1968, in Philadelphia, Denver, and Washington, D.C., and at Los Angeles, May through October, 1967

on the precursor concentrations of hydrocarbons and NO_x but also on their ratio. Thus, if NO_x is at 80 $\mu gm/m^3$ the same maximum possible oxidant level, 200 $\mu gm/m^3$, may occur if hydrocarbons range from 200 to 930 $\mu gm/m^3$. Conversely, if nonmethane hydrocarbons are at 200 $\mu gm/m^3$, NO_x levels may range from 80 to 320 $\mu gm/m^3$. There seems to be an optimum ratio of hydrocarbons to NO_x which yields the maximum attainable oxidant levels.

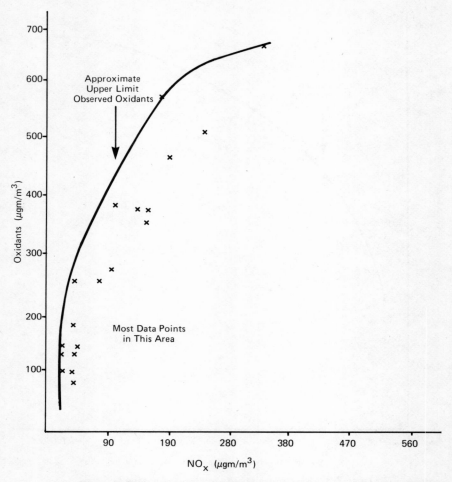

Figure 7-9. Maximum One-Hour Average Oxidant Concentrations as a Function of 6:00 to 9:00 a.m. Averages of Total Nitrogen Oxides in Washington, D.C., June through September, 1966 through 1968, and in Philadelphia and Denver, June through September, 1965 through 1968

The shape of the curves also indicates that reductions in hydrocarbons are more important than decreases in NO_x. "At practically every HC-NO_x concentration, reductions in HC without NO_x reduction result in a reduction in O_3, whereas NO_x reduction without HC reduction does not always lead to a reduction in O_3" [Environmental Protection Agency 1971a, p. 4-13]. Consequently, the criteria document for NO_x concludes, "The data base suggests that reductions

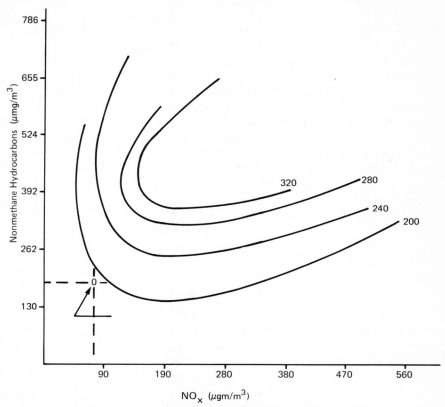

Figure 7-10. Approximate Isopleths for Selected Upper-Limit Maximum Daily One-Hour Average Oxidant Concentrations as a Function af the 6:00 to 9:00 a.m. Averages of Nonmethane Hydrocarbons and Total Nitrogen Oxides in Philadelphia, Washington, D.C., and Denver, June through August, 1966 through 1968

in HC should be the primary step for control of oxidants. Coupled with HC control, NO_x must be controlled at a level that will hold ambient NO_2 values below the level of adverse health effects" [Environmental Protection Agency 1971a, p. 4-20].

This data base is limited, however. It consists of the upper-limit points, about one percent of the 125 days of measurements. All these measurements were taken primarily between June and September of four years, in three cities, with air quality levels measured at one station in each city. The oxidant levels cited are the one-hour daily maximums, recorded on days when atmospheric conditions—low wind speeds, high temperatures, intense sunlight, and surface inversions—resulted in maximum oxidant potential.

Analysis of observational data on maximum limits provided the federal analysts with the DAQ for hydrocarbons: a maximum 6:00 for 9:00 a.m. average of 125 μgm/m^3. Barth set this limit in order to prevent the occurrence of a peak oxidant level of 250 μgm/m^3 (see Figure 7-9). The low HC level was determined by extrapolating from about ten data measurements in the lowest range, where instrument difficulties are common. Barth writes that "the drawing of quantitative conclusions from this graph is not wholly justified" [Barth et al. 1970, p. 1640]. Nevertheless, in the absence of better data, the analysis does draw such conclusions.

In order to calculate the desired reduction in auto HC emissions, the federal analysts used the highest nonmethane HC reading in 1967—3,500 μgm/m^3, which occurred in Los Angeles. Other data consisted of a background level of 100 μgm/m^3 and an average 1967 emission rate of 14.84 gm/mi. Calculations were:

$$\text{Percentage Reduction} = \frac{(2.18 \times 3{,}500) - 125}{(2.18 \times 3{,}500) - 100} = 99\%$$

$$14.84 \times (100\% - 99\%) = .15 \text{ gm/mi average allowable nonmethane HC emissions.}$$

Thus, the analysis concludes that a 99 percent reduction of nonmethane hydrocarbons from 1967 auto exhaust emissions levels is needed in 1980 to insure that a peak oxidant level of 250 μgm/m^3 is not reached in 1990.

To lower NO_x levels enough to guarantee that oxidants could not exceed 125 μgm/m^3 for an hour, a goal of 49 μgm/m^3 NO_x would have to be set (see Figure 7-10). But the federal report estimates that no known device could reduce auto emissions far enough to achieve that level. Consequently, the report recommends control of hydrocarbons to bring down oxidant levels, but only suggests limits on NO_x emissions sufficient to prevent any adverse health effects from nitrogen dioxide (NO_2) levels. The federal analysis makes the conservative assumption that in Los Angeles, the city of maximum health risk from auto air pollution, all NO_x is NO_2. This may occur at midday on days with intense sunlight. But most of the time NO is the major form of NO_x in urban atmospheres [Environmental Protection Agency 1971a].

The NO_x goal for DAQ of 190 μgm/m^3 for a one-hour average was taken from the then preliminary NO_x criteria document. This level is near the concentration associated with increased rates of respiratory disease in children as reported in the Chattanooga TNT plant area study [Environmental Protection Agency 1971a] although that study reports a twenty-four hour mean over a period of six months, not a one-hour average. In addition, proximity to a TNT plant is an unusual situation, so the results may not be generalizable to auto-produced NO_2. Unfortunately, however, the Chattanooga epidemiologic studies of NO_2 effects are the only ones available.

The maximum 1967 Los Angeles NO_2 reading for one hour was taken as 1,300 $\mu gm/m^3$. A background level of 8 $\mu gm/m^3$ and a 1967 average NO_x emission rate from cars of 5.93 gm/mi are also part of these calculations:

$$\text{Percentage Reduction} = \frac{(2.18 \times 1300) - 190}{(2.18 \times 1300) - 8} = 93.6\%$$

5.93 x 6.4% = .38 gm/mi average allowable NO_x emissions.

Thus, the report concludes that a 93.6 percent reduction from 1967 exhaust emissions of NO_x is needed in 1980 to ensure that an NO_2 level of 190 $\mu gm/m^3$ is not exceeded for an hour in 1990.

The authors of the article, all officials with the EPA responsible for advising decision-makers on effects of air pollution, incorporate many assumptions and simplifications that provide large margins of safety to protect public health. Their analysis implies that such safety margins, used to assure that there will be no health damage, are worth the cost. The Clean Air Act Amendments of 1970 state that standards are to be set by the EPA to protect public health; they do not permit other factors, such as cost, to be taken into account [*Federal Register 36* (April 30) 1971].

Relation to Current Ambient Air Quality Standards

On April 30, 1971, pursuant to the Clean Air Act, the EPA promulgated national primary air quality standards [*Federal Register 36* 1971]. A comparison of these standards and the federal report's "desired air quality" shows them to be similar.

Pollutant	National Standard	*Desired Quality in the Federal Analysis Calculations*
CO	10 mgm/m³ max 8-hr avg	same
Oxidants	160 μgm/m³ max 1-hr avg	125 μgm/m³ max 1-hr avg
HC	160 μgm/m³ max 6-9 a.m. avg	125 μgm/m³ max 6-9 a.m. avg
NO_x	100 μgm/m³ annual arithmetic mean	190 μgm/m³ 1-hr avg

And on July 2, 1972, the EPA promulgated auto emission standards, in grams per mile, based on a 90 percent reduction from 1970 average emissions for HC and CO and from 1971 average emissions for NO_x [*Federal Register 36* 1971]. The standards differ from the results of the federal report, which based its calculations on 1967 average emissions and recommended some percentage reductions greater than 90 percent. Furthermore, the emissions testing procedure in effect at the time of publication of the Barth article produced numbers lower than the revised federal test procedure promulgated with the national standards:

Emission	National Standard	Federal Result
HC	.41 gm/mi (by 1975)	.15 gm/mi
CO	3.4 gm/mi	6.16 gm/mi
NO_x	.4 gm/mi	.38 gm/mi

These standards constitute current national policy.

Chapter Eight

Measuring the Value of Emissions Reductions

William R. Ahern, Jr.

Few attempts have been made to estimate the benefits that might accrue from reductions in air pollution [Lave and Seskin 1971]. The favored approach has been to estimate the total cost of a category of damage that might be caused partly by air pollution. For example, one might estimate the total economic cost of chronic bronchitis [Ridker 1967] or of methods used to protect steel from corrosion and then attribute a percentage of this total cost to air pollution. The benefit to be obtained from air pollution reduction would be this percentage times the total cost. One study by Ridker [1967] estimates the total economic costs of seven disease categories and then estimates that 18 to 20 percent of the total cost of the diseases is due to air pollution. The .18 to .20 coefficients are derived from two studies that compare respiratory and lung cancer mortality rates of rural and urban populations. Ridker's study is primarily an exposition of measurement techniques, so the measurement of the total cost of the disease categories is detailed and complex. The calculations attributing 20 percent of this damage to air pollution, however, are much less precise.

A series of studies by Lave and Seskin [1970, 1971] uses multiple regression techniques to "explain" the variance in mortality rates among 117 Standard Metropolitan Statistical Areas (SMSA's) for 1960. The only two pollutants considered are sulfur oxides and particulates. None of Lave and Seskin's work deals with the pollutants associated with auto emissions. The main variables included in the regressions are racial composition, income, density, and particulate and sulfate levels. The regression coefficients estimated for the two air pollution variables are then used to predict possible changes in mortality rates when air pollution is reduced by 50 percent. The levels of the two pollutants with statistically significant coefficients are the minimum biweekly readings [1971].

One of Lave and Seskin's studies yielded the frequently quoted figure that the value of the estimated health losses from air pollution in 1968 totaled

$6.06 billion [1970]. The study used estimates for 1963 of the total costs of cardiovascular illness, respiratory disease, and cancer. The analysts then chose estimated percentages of these costs that might have been prevented by a 50 percent reduction in air pollution. These percentages were not derived quantitatively but were guesses based on some studies showing possible relationships between air pollution and the diseases [1970]. The values chosen were:

Disease Category	Percentage Attributed to Top 50 Percent Air Pollution	Total Disease Cost in $ Billions	Air Pollution Cost in $ Billions
Cardiovascular	10	4.68	.47
Respiratory	25	4.88	1.22
Cancer	15	2.60	.39
Total			2.08

Thus, the study suggests that the total annual saving from a 50 percent reduction in air pollution would be $2.08 billion.

An EPA study [Barrett and Waddell 1972] summarizes research on the costs of air pollution and then estimates that a 100 percent reduction in air pollution would save double the amount suggested by Lave and Seskin, or $4.16 billion. As this represents 0.7 percent of the 1963 Gross National Product (GNP) of $590.5 billion, the same percentage of the 1968 GNP of $865 billion is $6.06 billion. The analysis that originally produced this figure has certain deficiences, however. A major one is that the percentages of the costs of the three disease categories attributed to the top 50 percent of air pollution are not founded on medical evidence that establishes a causal relationship, but rather on the subjective estimates of Lave and Seskin, who are economists. And they seem to have had strong prior beliefs that air pollution causes significant health damage [McKean 1968].

Further, they contend that 25 percent of the cost of all respiratory disease in this country could be prevented by a 50 percent reduction in air pollution. If this were true, people living in rural areas would have significantly fewer such ailments than those living in urban areas. But national health statistics indicate that SMSA populations have 194 acute health conditions per 100 persons per year, whereas non-SMSA, nonfarm populations, which are not exposed to air pollution, have 188 per 100 persons [U.S. Department of Health, Education, and Welfare 1970a]. This is a small difference. Thus, attribution of such large percentages of the costs of cardiovascular and respiratory illness and cancer to a single factor, air pollution, is not warranted. More research is needed to determine the magnitude of air pollution's contribution to these illnesses.

Members of the EPA's Air Pollution Control Office report that

> Lave and Seskin seem to have a stronger faith in the magnitude, sign and statistical significance of their regression coefficients than what their analysis would seem to support. Their many statements about the causes of these "effects" are not as justified as they seem to conclude. Yet in fairness, despite the author's questionable data, analysis, and extended discussion of their results, their final cost estimate is believed to be reasonable. [Barrett and Waddell 1972, p. I-J12].

The cost estimate has been frequently cited, quite naturally, because it is the result of the only major attempt to carry out a difficult and complex estimation procedure. The reasonableness of the estimate, however, is still open to question. The methods used by Lave and Seskin to estimate the total cost of a damage category and to assign a percentage of that cost to air pollution leave out steps in the descriptive model (see Figure 7-1). Moreover, questions about the sources of emissions, exposure, the character of the damage, and the separate valuation of the damage attributed to air pollution are not posed. A new method is needed, therefore, both to deal with these questions and to provide estimates of marginal costs as well as total costs of air pollution.

AN APPROACH FOR VALUING REDUCTIONS IN AUTO EMISSIONS

The approach to benefit evaluation used here differs from the previously mentioned efforts to estimate air pollution costs in three major ways. The first is that instead of estimating the cost of a disease and then attributing part of that cost to air pollution, our approach uses a common public health measuring rod (a day of restricted activity) for estimating the impact of air pollution on each disease category. Thus the adverse health effect is measured first, and then it is valued. Second, the estimation of health effects is made by medical experts in the air pollution field. The third major difference is that we attempt to measure and value various percentage reductions in auto emissions. Such information about the value of increments in auto emissions reduction is needed to evaluate the benefits that may be foregone under policy options such as delaying or relaxing auto emission standards.

Our approach progresses systematically through the steps in the descriptive model of health damages presented in Figure 7-1. At many points simplifications and assumptions must be made because of lack of data and resources. But throughout the discussion we try to show how improved data can be incorporated into the approach in order to produce more refined and accurate results. Note that this valuation approach is only applied to the short-run effects of air pollution on health, although it is applicable to other phenomena as well. Finally,

178 Clearing the Air

the goal of the approach is to make a national estimate of the value of auto emissions reduction, not an estimate for an urban area or a state. The approach and a limited attempt to implement it are described in the remaining pages of this section. Figure 8-1 is an outline of the approach; it summarizes the steps and shows how they relate to the descriptive model of the valuation process.

Process	Steps
Auto Emissions (AE)	1. Equate auto's contribution to air pollution over urban areas with auto's contribution, by weight, of each pollutant.
Pollutant Concentrations in the Ambient Air (AA)	2a. Represent pollutant concentrations by constructing a representative urban area, using mean levels for eight cities. Indicate hours per year during which pollutant levels which would affect health are exceeded. 2b. Project future AA for different reductions in auto emissions.
Exposure (EX)	3. Disaggregate population by place of exposure—central city and suburbia.
Damage to Health (D)	4a. Disaggregate population into categories of susceptibility to adverse health effects from air pollutants. 4b. Develop a common measuring rod for estimating damage (D), the EDRA. 4c. Estimate the damage (D) in EDRA's attributed to exposure to AA for each category of susceptibility. 4d. Aggregate the amounts of damage for different levels of emissions reduction.
Value (V)	5. Estimate the dollar value of the damage associated with different levels of emissions reduction from the standpoint of the people affected.

Figure 8-1. Outline of an Approach for Valuing Reductions in Auto Emissions

Step 1: The Automobile's Contribution to Air Pollution

Auto emissions contribute to ambient levels of pollutants that are of concern to health officials—carbon monoxide and photochemical oxidants. This discussion is designed to answer the question, "What part of national ambient CO and photochemical oxidant concentrations can be attributed to auto emissions?" Ambient air pollutant levels at any time in any place are a function of auto emissions and other emissions (both natural and man-made), topography, temperature, wind, sunlight, energy, and other factors. Thus, answering the question requires a highly complex mathematical model.[1]

The model used here is more complex than that used by Barth, but simpler than that needed for regional or local problems. It is useful mainly for national policy-making purposes. The model takes the percentage, by weight, of CO emissions contributed by autos and uses that percentage as the auto's contribution to ambient CO levels. Emissions data analyzed by HEW [1970a] show that this percentage ranges from 60 percent in Philadelphia to 99 percent in Washington, D.C., for all transportation sources. The mean of eight CAMP cities is 87 percent. Estimates of sources of CO emissions [U.S. Department of Health, Education, and Welfare 1970a] indicate that in urban areas, about one-seventh of CO emissions from transportation originate from heavy-duty vehicles. Therefore, using this model to arrive at a rough national approximation, six-sevenths of 87 percent or about 75 percent of ambient CO concentrations over large urban areas may be attributed to automobile emissions.

This model cannot be used for photochemical oxidants since the precursor emissions, hydrocarbons and oxides of nitrogen, are different from the ambient air pollutant. Emissions estimates indicate that, nationally, about half of all NO_x and HC emissions in urban areas come from transportation sources. But the cities with weather conditions highly favorable for photochemical oxidant formation, especially on the southern California coast, have comparatively few industrial sources and limited space heating sources of NO_x and hydrocarbons. Since this chapter is concerned with oxidant levels high enough to affect health, and since these occur primarily in cities, the gross simplification is made here that 100 percent of photochemical oxidant levels may be attributed to auto emissions of NO_x and hydrocarbons. This overstates the auto's role in oxidant formation. But most of the cities with the worst oxidant smog—Los Angeles, Pasadena, San Diego, Denver, Sacramento—have comparatively few heavy industrial sources of NO_x and hydrocarbons (see Table 8-2).

Step 2: Pollutant Concentrations in the Ambient Air for Various Reductions in Auto Emissions

Levels of CO and photochemical oxidants high enough to irritate and/or affect health occur primarily over large urban areas. Small cities have fewer vehicles and more frequent ventilation with clean air. Here, we define urban areas as those with 250,000 or more people. Locations outside these areas

---

[1] Much effort has gone into attempts to develop complex models that can quantitatively estimate the effect of weather on the dispersion of pollutants. Sections of the air quality criteria documents on pollutant levels in the ambient air cite some of these. Current research on this and the other air pollution topics is listed in the annual *Guide to Research in Air Pollution* published by the EPA and compiled by the Center for Air Environment Studies of Pennsylvania State University [1971].

are considered nonurban. Ambient pollutant levels vary widely over these areas. Figure 8-2 shows CO levels over Los Angeles. The levels are high over the central

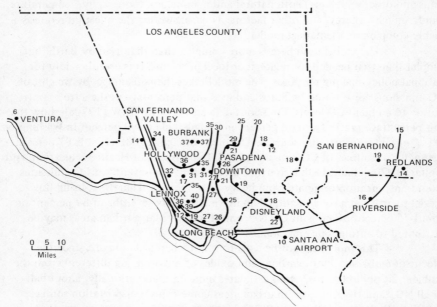

Figure 8-2. Eight-Hour Averaging Time Carbon Monoxide Concentrations (ppm) Exceeded 0.1 Percent of the Time in the Los Angeles Area, 1956 Through 1967

Note: From U.S. Department of Health, Education, and Welfare [1970a]

city and other areas with high traffic density, and they decrease with distance from these areas. Many air monitoring stations are needed to measure air pollution levels accurately in order to answer the question, "For what time period and to what CO and oxidant concentrations are people in these urban areas exposed?" Though monitoring data for 1967 are limited, this year was chosen because it was used as the base year in the federal analysis of auto emissions reduction and because it predates most auto emissions controls.

    A rough approximation of average urban CO and oxidant levels can be formed with data from the eight cities in CAMP. The cities (Los Angeles, San Francisco, Chicago, St. Louis, Cincinnati, Philadelphia, Washington, D.C., and Denver) each had a monitoring station, usually in or near the downtown area. All are large cities with many autos, so using their air pollution readings to compute an average for all urban areas creates a bias toward high levels. Data from other cities are now being gathered by the EPA as part of a national air sampling program.

Carbon Monoxide. The CAMP data for CO are presented in Table 7-1. The measurements used are for eight-hour averaging periods and are presented as the percent of time in the year that a concentration is exceeded. For example, at Denver's CAMP station an eight average CO level of 12 ppm (10.5 mgm/m^3) was exceeded 10 percent of the time. The goal here, however, is not to represent a city's ambient level but to get a rough estimate of aggregate national urban levels. We do this by averaging the CO concentrations of the eight cities for each percentage of time, 0.1 to 90 percent. Converting these percentages into hours of the year gives the data presented in Table 8-1 for a representative CAMP station. Only concentrations above the national standard are considered since, on the basis of current knowledge, lower levels do not cause adverse health effects.

Table 8-1. Distribution of Carbon Monoxide Levels for a Representative CAMP Station in 1967 (Only Hours in Which the National Standard Is Exceeded Are Shown)

Range of Eight-Hour Average CO Readings in mgm/m^3 (ppm)	Approximate Number of Total Hours in the Year that Fell Within Averaging Periods for Which Observed Concentrations Were Present
10 to 10.5 (11.5 to 12)	600
10.5 to 15.5 (12 to 18)	800
15.5 to 21 (18 to 24)	80
21 to 34 (24 to 39)	10

Since the CAMP stations are downtown, their readings do not represent the entire urban areas. Here, however, the representative urban area is disaggregated into two parts, the central city and the suburbs. The central city is the political jurisdiction that forms the core city of an SMSA, as the city of Boston is the central city of the Boston SMSA. The other parts of the SMSA are considered the suburbs. The CO levels, as indicated by Figure 8-2, can be expected to be lower in the suburbs. Statistical analysis of CO aerometric data indicates that the central urban areas have about twice the CO levels of residential areas; since central cities contain some residential areas, a factor of one-third is used to convert the CAMP readings into approximate suburban CO levels for the representative urban area.

The next step is to predict what ambient CO levels for the representative urban area would result from different percentage reductions in auto emissions. Reductions of 50 percent, 75 percent, and 90 percent are considered. To make such predictions accurately would require the treatment of many complex and uncertain variables, including future auto population growth, urban density and area changes, and meterological changes. A simple predictive model for air quality is presented in an article published by the Environmental Quality Laboratory of the California Institute of Technology [Lees et al. 1972]. It has been adapted here for use with CO levels over the representative urban area. It assumes that for the same meterological conditions, if the level of emissions is reduced by a certain percentage, the ambient concentrations are also reduced by the same percentage. Thus, if CO emissions are reduced 50 percent, then the number of hours in a year for which a particular eight-hour average CO concentration is exceeded is the same as the number of hours per year for which twice this concentration was exceeded at twice the emissions level (see the horizontal dashed line in Figure 8-3). Accordingly, if CO emissions were reduced by 50 percent (the dashed curve), given 1967 conditions, the national standard would be exceeded during approximately twenty-five hours over the year.

But air pollution policy is concerned with the future. A 50 percent

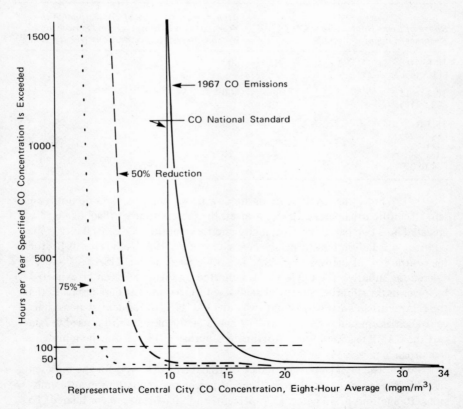

Figure 8-3. Effect of Reductions in 1967 CO Emissions on CO Levels for a Representative Central City

reduction in emissions per car would not result in a 50 percent reduction in ambient CO twenty years or so from 1967 because of growth in the number of auto miles driven. The EPA calculated that the number of autos in the nation by 1990 will be 2.18 times greater than that in 1967. The growth in emissions, over a constant area, would probably be less than that. An emissions growth factor of 1.5 for a constant area is accurate enough for use here.

In 1990, then, a 50 percent reduction in CO emissions per car and a 1.5 emissions growth rate would yield, according to the model, a 25 percent reduction in ambient CO levels. Likewise, a 75 percent reduction in CO emissions would result in a three-eighths reduction of 1967 levels. The ambient CO levels for a representative central city predicted by the calculations are shown in Figure 8-4. Three lines represent the estimates of the concentration and duration of CO to which national central city populations are exposed. CO emissions from other sources are considered reduced by the same percentage as auto CO emissions. This assumption and the calculations can be varied for different policies on CO emissions from stationary

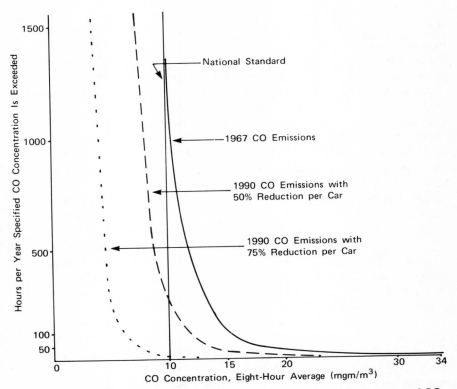

Figure 8-4. Effect of 50 Percent and 75 Percent Reductions of CO Emissions per Car on CO Levels for a Representative Central City in 1990.

sources. The 90 percent reduction lines are not shown since they result in no hours above the national standard. One-third of 1967 levels is used to represent suburban CO levels. This results in approximately ten hours during which the national standard is exceeded. The maximum suburban CO level would be one-third of 34, or 11.3 mgm/m^3. If emissions were reduced 50 or 75 percent, suburban CO levels would then result in no hours above the national standard.

Photochemical Oxidants. The same procedure, with a few modifications, is used to estimate the concentrations and duration of photochemical oxidants to which central city and suburban residents of a representative urban area are exposed. Table 8-2 presents the CAMP data for the eight cities for which readings are averaged to construct a cumulative frequency distribution for a representative urban area. Converting the percent of hours into number of hours yearly and the parts per million to micrograms per cubic meter (μgm/m^3) gives the concentration and duration data in Table 8-3.

Particular photochemical oxidant concentrations can result from a wide range of precursor NO_x and HC emissions. A 50 percent reduction in reactive HC emissions accompanied by a smaller or no NO_x reduction therefore may not result in a 50 percent reduction in oxidant levels. Estimating the combinations of NO_x and HC emission reductions that will result in selected percentage ambient oxidant level reductions is a highly complex and uncertain process. Reductions might be obtained by reducing NO_x or HC emissions alone. Here the air quality prediction model assumes that 50 and 75 percent reductions in NO_x and HC emissions result in, respectively, one-fourth and five-eighths reductions in ambient oxidants. This assumes a linear relation between oxidants and precursor emissions, an unlikely assumption given the complex and, at this time, controversial but definitely nonlinear nature of the relationships (see Figures 7-8 through 7-10). Linearity is assumed for simplicity and for lack of anything more precise. The cumulative frequency distribution of oxidant concentrations is graphed in Figure 8-5. A 90 percent reduction in oxidant precursor emissions is expected to result in no hours exceeding the national standard.

Figure 8-5 shows the concentration and duration of photochemical oxidants for a representative central city. Since photochemical oxidants form gradually in an air mass, the ambient concentrations can be high not only over central cities (and areas of high traffic density) but also over downwind portions of the suburbs. Thus Riverside, downwind of Los Angeles, receives at 4:00 p.m. the oxidants generated over Los Angeles at 11:00 a.m. [U.S. Department of Health, Education, and Welfare 1969a]. We assume that one-third of the suburbs of the representative urban area are exposed to the same oxidant levels as the central city and that the remaining two-thirds of the suburbs experience low or negligible oxidant levels (see Figure 8-6).

Step 3: Exposure
Exposure is a function of pollutant concentration and duration.

Table 8-2. Cumulative Frequency Distribution of Hourly Average Oxidant Concentrations in Selected Cities, 1964-1965

City	Percent of Hours with Concentrations Equal to or Greater than Stated Concentrations (ppm)									1964-1965 Yearly Average (ppm)
	90	70	50	30	10	5	2	1		
Los Angeles	0.01	0.01	0.02	0.04	0.10	0.14	0.18	0.22		0.036
Denver[a]	0.01	0.02	0.03	0.04	0.06	0.08	0.10	0.12		0.036
St. Louis	0.01	0.02	0.03	0.04	0.06	0.07	0.09	0.11		0.031
Philadelphia	0.01	0.02	0.02	0.03	0.06	0.08	0.11	0.14		0.026
Cincinnati	0.01	0.02	0.02	0.04	0.06	0.07	0.08	0.10		0.030
Washington, D.C.	0.01	0.01	0.02	0.03	0.06	0.07	0.09	0.10		0.029
San Fransisco	0.01	0.01	0.02	0.03	0.04	0.05	0.06	0.07		0.019
Chicago	0.01	0.01	0.02	0.03	0.05	0.06	0.08	0.08		0.028

[a]Eleven months of data beginning February 1965.
Note: From U.S. Department of Health, Education, and Welfare [1969a]

Table 8-3. Distribution of Oxidant Levels for a Representative CAMP Station in 1967 (Only Hours During Which the National Standard Is Exceeded Are Included)

Range of One-Hour Average Oxidant Reading in $\mu gm/m^3$ (pphm)*	Approximate Number of Total Hours in the Year in Which Levels in the Specified Range Are Present
160 to 200 (8 to 10)	175
200 to 240 (10 to 12)	90
240 to 540 (12 to 27)	80
540 to 1160 (27 to 58)	10

*parts per hundred million

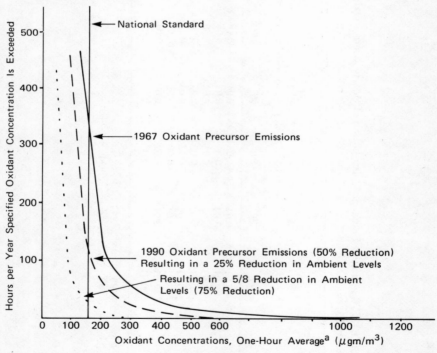

Figure 8-5. 1967 Oxidant Concentrations and the Effects, by 1990, of 50 Percent and 75 Percent Reductions in Oxidant Precursor Emissions (HC and/or NO_x) for a Representative Central City and One-Third of the Suburban Area

[a] These hours are usually between 10:00 a.m. and 2:00 p.m. daily.

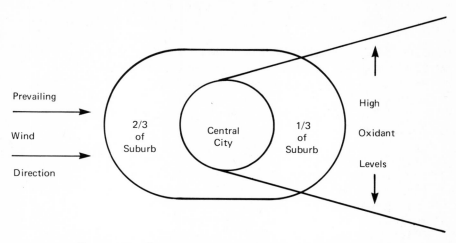

Figure 8-6. Oxidant Transport over a Representative Urban Area

In the preceding step the concentration and duration of CO and oxidants were estimated for two geographical categories—the central city and the suburbs of SMSA's with a population of more than 250,000. This step addresses the problem of estimating the number of people exposed to these pollutant concentrations. Data from the *Statistical Abstract of the United States* [U.S. Bureau of the Census 1968] and from the *New York Times Encyclopedic Almanac* [1970] show that approximately the following percentages of the 1967 national population were located in the two categories: Central city—26 percent; suburban—32 percent. The other 42 percent lived outside SMSA's with a population larger than 250,000. Thus, we estimate that 26 percent of the population is exposed to the representative central city CO concentrations of Figure 8-5, 32 percent to the representative suburban CO concentrations, and 26 percent plus 11 percent (one-third of 32 percent), or 37 percent, to the oxidant concentrations of Figure 8-6.[2]

Step 4: Damage to Health

Categories of Susceptibility to Adverse Health Effects. Carbon monoxide and photochemical oxidant exposure affects different people in different ways. To identify and measure adverse health effects attributable to the pollutant exposure estimates just presented, we must disaggregate the general population into broad categories of susceptibility to such adverse effects.

---

[2]These are first-level disaggregations. More and better data on population, ambient levels, population movement, and other factors, plus time and research resources, could yield more accurate research estimates.

Three such categories are:

1. Those susceptible because of *chronic* adverse health conditions.
2. Those susceptible because of *acute* adverse health conditions
3. The healthy (those not in categories 1 and 2).

Chronic conditions may be further divided into the following more specific categories[3]:

1. Asthma and hay fever
2. Sinusitis and bronchitis
3. Other respiratory conditions
4. Heart conditions
5. Symptoms referable to the respiratory system
6. Old age (here defined as over 65).

Data on the numbers of people suffering these chronic conditions are available from the Public Health Service's series, *Vital and Health Statistics* [1971b]. One issue tells how many people have chronic conditions that prevent them from engaging in certain degrees of physical activity. Activity limitations range from inability to participate in sports to inability to hold a job. Unpublished data available at the National Center for Health Statistics report how many people have these chronic conditions but are able to lead normal lives.[4] They are especially susceptible to health damage from air pollutant exposure since, despite their conditions, they move about freely in the ambient air. One problem in using these categories is that it is difficult to estimate the degree of overlap among them. For example, how many aged people also have one or more of these chronic conditions?

The second large category consists of those who might be susceptible to air pollution because of an acute condition—one lasting less than three months. Data on these conditions also are published by the Public Health Service [1971a]. The set of illnesses within this second category includes:

1. Upper respiratory conditions (common cold and others)
2. Influenza
3. Other respiratory conditions (pneumonia, acute bronchitis, others).

Not only the number of cases per year but also the number of days of restricted

---

[3]Cancer is not included because no epidemiological evidence indicates that it is affected by CO or oxidants. This set may, of course, be expanded or contracted as more research results become available.

[4]These data were provided by Mr. Joel Kavet of the Harvard School of Public Health.

Table 8-6. Estimation of the Number of Days of Restricted Activity Attributable to Carbon Monoxide Levels for 1967 Levels and for 50 Percent and 75 Percent Reductions (Susceptibility Category: Persons (Cases) with Acute Conditions, Central City Exposure)

Condition	Millions of Cases per Year	Number of Days Restricted Activity	Estimate the Number of Days of Restricted Activity that you Attribute to Annual Levels of Carbon Monoxide as Displayed in Figure 8-4		
			1967 Levels (Solid Line)	50 Percent (Dashed Line)	75 Percent Dotted Line)
Upper respiratory conditions	38.0	77.6			
Influenza	15.0	50.5			
Other respiratory conditions	1.9	13.9			
Other (specify)					
Totals					

Table 8-7. Total National EDRA's Assigned by Three Estimators to Chronic and Acute Health Conditions for Three Pollutant Concentration Profiles of CO and Oxidants

	1967 Levels	1990 Level with 50 Percent Emissions Reduction	1990 Level with 75 Percent Emissions Reduction
Carbon monoxide			
Medical doctor	0	0	0
Epidemiologist	35,000	30,000	6,250
Occupational health expert	27,725	9,820	4,840
Oxidants			
Medical doctor	200	25	0
Epidemiologist	58,250	56,250	29,500
Occupational health expert	263,000	158,000	81,000

Note: The author was surprised by the low nature of the estimates. But his main source of information on health effects had been the summarizations of studies in the air quality criteria documents. See pp. 152-162 in Chapter 7 for problems with these summarizations.

mates. The figures on ambient pollution concentrations (Figures 8-4 and 8-5) provide estimates of the number of hours per year a person in the central city or suburb might be exposed to potentially harmful levels of CO or oxidants. The number of hours is represented on the vertical axes. Thus, for central city CO, a 1990 concentration of 10 mgm/m^3 is exceeded for about 250 hours with a 50 percent reduction per car in CO emissions (dashed line). Using 1967 data, we multiply the hours of damage per person by the number of persons in the central cities and suburbs exposed to the CO and oxidant levels. The result is the total number of hours people are exposed to such levels nationally. A subjective but reasonable conversion factor of 50 hours equal to one EDRA can be used to convert these hours into EDRA's. Thus, for CO, the high estimate might include all hours in eight-hour averaging periods during which CO exceeds the national standard of 10 mgm/m^3. The national standard is considered a high estimate because it incorporates many margins of safety designed to prevent *any* adverse health effects. Values for the high estimate are given in Table 8-8. The low estimate might use all hours during which CO exceeds 15.5 mgm/m^3. Total hours are given in Table 8-9.

A similar set of calculations can be made for oxidants using the concentrations represented in Figure 8-5. For the high estimate, all hours above 160 $\mu gm/m^3$ (national standard) are considered irritating, and those above 540 $\mu gm/m^3$ are judged especially damaging. These two levels are chosen because one experiment indicates that eye irritation may start at 200 $\mu gm/m^3$, whereas another [U.S. Department of Health, Education, and Welfare 1969a] indicates that nose and throat irritation may start at 540 $\mu gm/m^3$.

Table 8-8. High Annual National Health Damage Estimates Using Eight-Hour Averaging Period When CO Concentration Exceeds 10 mgm/m^3 [a]

		1967 Levels	1990 Levels with 50 Percent Emissions Reduction	1990 Levels with 75 Percent Emissions Reduction
Number of hours per person per year exposed to CO above 10 mgm/m^3	Central City	1,500	250	20
	Suburbs	10	0	0
Millions of hours of exposure per year (population x hrs/person)	Central city	70,500	11,750	940
	Suburbs	580	0	0
Millions of total EDRA's with conversion factor of 50 hrs/EDRA	Central city	1,410	235	19
	Suburbs	11.6	0	0
Millions of total EDRA's	Total	1,421	235	19

[a] Data are from Figure 8-4.

Table 8-9. Low Annual National Health Damage Estimates Using Eight-Hour Averaging When the CO Concentration Is 15.5 mgm/m^3 or Higher (Figures Given for the Central City Only)

	1967 Levels	1990 Levels with 50 Percent Emissions Reduction	1990 Levels with 75 Percent Emissions Reduction
Total hours per person per year exposed to 15.5 mgm/m^3 or above	100	25	0
Millions of person-hours per year	4,700	1,170	0
Millions of total EDRA's using conversion factor of 50 hours = 1 EDRA	94	23.5	0

Using a larger conversion factor to EDRA's (fifty hours equal to one EDRA) for levels of 160 to 540 $\mu gm/m^3$ than for levels of 540 $\mu gm/m^3$ and up (twenty hours equal to one EDRA) reflects the relative damage inflicted by the different levels. See Table 8-10 for a presentation of the results. Table 8-11 presents low estimates, using only hours above 540 $\mu gm/m^3$ with a conversion factor of forty hours equal to one EDRA. Aggregating the EDRA estimates for each pollutant gives the results in Table 8-12. The figures in Table 8-12 show irritant and discomforting effects of the two pollutants vastly outweigh the effects which exacerbate already existing illness. A comparison of the high and low estimates for the healthy population reveals a very wide range between them as well. The total benefit functions for reductions in the two pollutants are presented in Figure 8-7.

Policy Implications. The value of reductions in auto emissions can now be measured in prevented EDRA's. The increments in the numbers of EDRA's prevented annually by each 1 percent reduction in auto CO and oxidant precursor emissions are shown in Figure 8-8. These marginal benefit functions are derived from the two total benefit functions (see Figure 8-7). Looking at the high CO estimate, roughly 25 million EDRA's can be prevented by each 1 percent reduction in CO emissions up to a total 50 percent reduction. After that, the marginal benefit, in prevented EDRA's, declines rapidly and reaches zero before a total 90 percent CO emissions reduction.[8] If the low estimate is more accurate, only about 1 or 2 million EDRA's can be prevented annually by each 1 percent reduction in CO emissions up to a total 75 percent reduction, at which point the marginal benefit is zero.

The damage to health from photochemical oxidants is about 20 percent of that from CO. One percent reductions in oxidant precursor emissions might prevent from 1 million (low estimate) to 5 million (high estimate) annual EDRA's. That oxidants, the major components of smog, are less damaging than CO may seem intuitively wrong. It must be emphasized that the only category of damage being considered here is short-run damage to health. Carbon monoxide is a colorless and odorless gas that does not corrode materials or affect visibility. Oxidants, however, are visible and offensive to smell, and some adversely affect materials and vegetation. Thus, reducing oxidant precursor emissions "buys" more than prevented EDRA's. Health effects, however, are the major concern of an emissions reductions policy.

A policy that imposes a cost on the public to achieve a reduction in auto emissions implicitly values the EDRA's prevented by the reduction. For example, if reducing auto CO emissions from 50 to 75 percent of 1967 levels costs, nationally, $1 billion a year, and, according to the high estimate (see Figure 8-7),

---

[8]A 90 percent emission reduction from 1967 levels is considered to yield all zeroes. There is the possibility, however, that even with this reduction Los Angeles would experience a number of hours above the oxidant national standard.

Table 8-10. High National Oxidant Damage Estimates[a]

	Levels of Oxidants ($\mu gm/m^3$)	1967 Levels	1990 Levels with 50 Percent Emissions Reduction	1990 Levels with 75 Percent Emissions Reductions
Number of hours per year a person is exposed	160-540	350	130	20
	540+	10	2	0
Missions of person-hours of exposure[b]	160	23,000	8,580	1,320
	540+	660	132	0
Millions of EDRA's, using 50 hrs = 1 EDRA for 160-540 $\mu gm/m^3$		462	176	26.4
Millions of EDRA's, using 20 hrs = 1 EDRA for 540+ $\mu gm/m^3$		33	6.6	0
Total EDRA's in millions		495	183	26.4

[a]Data are from Figure 8-5.

[b]Hours per person multiplied by central city population of 47,000,000 plus one-third of suburban population of 58,000,000.

Table 8-11. Low National Oxidant Damage Estimates[a]

	1967 Levels	1990 Levels with 50 Percent Emissions Reduction	1990 Levels with 75 Percent Emissions Reduction
Number of hours per year a person is exposed to 540 $\mu gm/m^3$	10	2	0
Millions of person-hours of exposure	660	132	0
Millions of EDRA's using 40 hrs = 1 EDRA conversion factor	16.5	3.3	0

[a]Data are from Figure 8-5.

198 Clearing the Air

Table 8-12. Total High and Low EDRA Estimates for CO and Oxidants, in Millions of EDRA's per Year

	1967 Levels	1990 Levels with 50 Percent Emissions Reduction	1990 Levels with 75 Percent Emissions Reduction
Carbon Monoxide			
Low[a] Chronic and acute conditions	0	0	0
Effects on healthy	94	23.5	0
Total	94	23.5	0
High[a] Chronic and acute conditions	.035	.030	.006
Effects on healthy	1,420	235	19.0
Total	1,420	235	19.0
Photochemical Oxidants			
Low[a] Chronic and acute conditions	.0002	.000025	0
Effects on healthy	16.5	3.3	0
Total	16.5	3.3	0
High[a] Chronic and acute conditions	.263	.158	.081
Effects on healthy	495	183	26.4
Total	495	183	26.5

[a]The lowest of the three experts' estimates is used as the low estimate, the highest as the high.

this reduction "buys" 220 million prevented EDRA's a year, then an EDRA is valued, in effect, at approximately $4.00. Policy-makers can then judge whether this value is too high, just right, or too low.

The graphs in Figures 8-7 and 8-8 can also be useful in judging other policy options. If a delay in emissions reduction is under consideration, the benefits from the delay can be compared with the cost of the number of EDRA's per year that will not be prevented. Similar comparisons can be made for changes in technology that may result in lower emissions over time.

Calculations of total EDRA's for any year beyond 1967 should be adjusted for increased population. If population climbs by 2 percent a year, then the number of total EDRA's for a year x years after 1967 should be multiplied by 1.02^x.

Step 5: Valuing Health Damage

The preceding section shows how EDRA's can be valued in the politi-

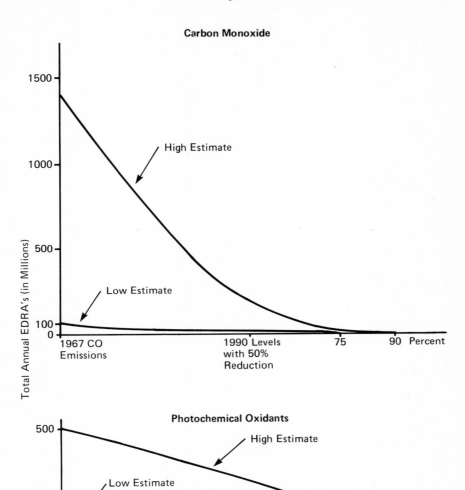

Figure 8-7. Graphs of Total Annual EDRA's Associated with Various Reductions in CO and Oxidant Precursor Emissions

cal process. Current national policy, as embodied in the Clean Air Act, explicitly judges the health damage prevented by reducing auto emissions from 85 to 90 percent to be worth whatever the cost of that reduction might be. In the future, however, the EPA and Congress may consider changes in the policy, such as relaxing the required percentage reduction, delaying emissions reduction requirements, or changing to another automotive technology. Consequently, it may

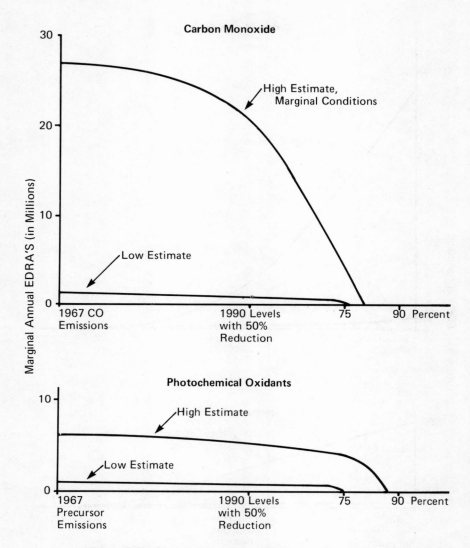

Figure 8-8. Marginal Annual EDRA's Attributed to 1 Percent Reductions in 1967 CO and Oxidant Precursor Emissions

still be worthwhile to attempt to value marginal reductions in auto emissions, especially from 75 to 90 percent, the range in which costs of reduction rise most rapidly.

Since costs are generally expressed in dollars, a further valuation step is used to answer the question: How much are x EDRA's worth in dollars? This requires estimation of a shadow price—a price for a commodity that is exchanged in an imperfect market or for which no market exists. There are a number of methods for estimating shadow prices [McKean 1968]. One is to use the price of other programs designed to achieve a similar outcome. The costs of public health programs, such as vaccination or immunization, might be used to determine what a "reasonable" cost of air pollution reduction might be. Another method is to leave the valuation to the political process. If the decision is made to reduce auto emissions by 50 percent, at a cost of $\$y$ that may prevent x EDRA's, then the political process has valued an average EDRA at $\$y/x$. This value might be checked for reasonableness and further adjustments in the policy made if the value does not appear to be reasonable. A third method is to use the market price of a commodity that delivers a comparable benefit. For example, a nasal decongestant might be considered to deliver relief nearly equivalent to that provided by preventing an EDRA. The market price could be adjusted for differences in the level of restricted activity prevented.

Such methods might be used, at least to determine the range of dollar values of a prevented EDRA. A further method is to estimate the willingness of various categories of the population to pay for preventing EDRA's. Intensive interviews [Schelling 1968] of samples of the population might be conducted to find out how much people are willing to pay for the benefits of various levels of auto emissions reduction.

Each of these methods has critical problems. At the present time, it is nearly impossible to accurately perceive, separate, and measure the health damage prevented by air pollution abatement. Strong feelings and misperceptions may sway estimates in widely different directions. Another problem is that several of the methods use current market values as a benchmark. But these market values only reflect the current distribution of income and the current preferences of the population; they do not anticipate conditions in the future, when the program may be operative, nor do they account for the costs of externalities.[9]

A way to deal with the most difficult of these problems is to estimate the willingness of parts of the population with different incomes to pay for prevented EDRA's. The uncertainties can be accounted for by choosing a wide range of EDRA valuations. To do this we first divide the population into income categories. Data for 1967 [U.S. Public Health Service 1971b] reveal the following breakdown of the national population according to family income:

Annual Income (Dollars)	Age 0-64 (Percent)	Age 65+ (Percent)
0-5,000	30	70
5,000-10,000	45	20
10,000 or more	25	10

---

[9] An externality is a by-product of a production process. The by-product imposes tangible or intangible costs on parties external to the process, costs not reflected in the price of the product since they are not counted as part of production costs. Air pollution is a classic externality.

The next step, to decide how much people in the income categories given above might be willing to pay to prevent an EDRA, is much more difficult. A range of values can be estimated by using proxy market commodities. (The main criterion here is that the values "sound reasonable.") For example, if a box of Alka Seltzer costs about $1.30 or a package of Contac $1.69, these prices indicate the low range of estimates of willingness to pay for the three categories of income[10]:

Income Category	Willing to Pay to Prevent 1 EDRA
$0-5,000	$0.50
$5-10,000	$1.00
$10,000 up	$2.00

The high values can be derived from the average cost of a visit to a doctor, $12:

Income Category	Willing to Pay to Prevent 1 EDRA
$0-5,000	$ 5
$5-10,000	$10
$10,000 up	$20

If the central city residents aged 0 to 64 suffer 216,000 EDRA's at 1967 levels of oxidants, then the formulas for the range of dollar values of that damage are:

income factor x total EDRA's x low willingness to pay

and

income factor x total EDRA's x high willingness to pay.

Thus, the total dollar values range from:

low = (30% x 216,000 x $0.50) + (45% x 216,000 x $1)
+ (25% x 216,000 x $2) = $240,400

to

high = (30% x 216,000 x $5) + (45% x 216,000 x $10)
+ (25% x 216,000 x $20) = $2,379,000.

Preventing those 216,000 EDRA's would be worth between $240,000 and $2,400,000.

Applying these willingness-to-pay factors for each income category to the total EDRA estimates of Table 8-12 gives the values in Table 8-13. Table 8-14 shows the incremental values of EDRA's prevented by reducing emissions from 1967 levels by 50, 75, and 90 percent. The data in Table 8-14 again demonstrate

---

[10]Remember that the wide variation in pain and suffering caused by air pollutants is considered in calculating EDRA's since an EDRA is an *equivalent* day of restricted activity, and conversion factors change various intensities of effects into the common measuring unit.

Table 8-13. High and Low Valuations of High and Low National Annual EDRA Estimates Attributed to CO and Oxidants[a], in $ Millions

		1967 Levels	1990 Levels with 50 Percent Emissions Reduction	1990 Levels with 75 Percent Emissions Reduction
Carbon monoxide				
Low EDRA estimate	Low valuation	102	26	0
	High valuation	1,020	260	0
High EDRA estimate	Low valuation	1,530	256	21
	High valuation	15,300	2,560	210
Oxidants				
Low EDRA estimate	Low valuation	18	3.6	0
	High valuation	180	36	0
High EDRA estimate	Low valuation	545	196	28
	High valuation	5,450	1,960	280

[a]Data are from Table 8-13.

the rapidly diminishing returns from increased auto emissions reduction, especially for the increment from 75 to 90 percent reduction.

RESEARCH ON HEALTH EFFECTS OF POLLUTION

The primary ongoing research project investigating the health effects of air pollution is the EPA's Community Health Effects Surveillance Studies (CHESS) program. William Ruckelshaus, former administrator of the EPA, described the program at Congressional Hearings on EPA appropriations for 1972:

> The CHESS program is a coordinated series of epidemiological studies designed to document the effects of changing ambient air on community health. . . . The CHESS concept is anchored to extensive environmental and health monitoring in sets of communities demonstrating an exposure gradient for specified pollutants. Three years of intense effort have been devoted to the design and field testing of the health impact indicators now a part of CHESS. The air quality aspects of CHESS will not be fully operational until 1973 and will evaluate the

Table 8-14. Incremental Values of National Annual EDRA's Prevented by Reducing 1967 Emissions by 50, 75, and 90 Percent, in $ Millions

		50 Percent Reduction ($ Millions)	50 to 75 Percent Reduction ($ Millions)	75 to 90 Percent Reduction ($ Millions)
Carbon monoxide				
Low EDRA estimate	Low valuation	76	26	0
	High valuation	760	260	0
High EDRA estimate	Low valuation	1,488	235	21
	High valuation	14,880	2,350	210
Oxidants				
Low EDRA estimate	Low valuation	14.4	3.6	0
	High valuation	144	36	0
High EDRA estimate	Low valuation	349	168	28
	High valuation	3,490	1,680	280

health effects information about the most important air pollutants. These effects will be continually monitored as pollution is controlled and can thus document the health benefits of abatement [U.S. House of Representatives, Subcommittee of the Committee on Appropriations 1971, pp. 354-55].

Thus CHESS will document health benefits to actual communities by reduced air pollutant emissions *after* the national emissions reduction policies have been implemented. This historical documentary approach is to be expected from the EPA since the Clean Air Act, which the EPA administers, requires a 90 percent reduction in auto emissions, regardless of marginal costs and benefits. This means, however, that no research data will be available from CHESS to help evaluate possible alterations in national policies. Of course, if future CHESS resul.s indicate, for example, that expensive policies yield little health improvement, then adjustments may be made on a post hoc basis. It is to the EPA's credit that the results of a federal program are being monitored at all. CHESS research so far has focused on the most toxic air pollutants, sulfur oxides and particulates, in five major urban areas. Its data will not be useful for evaluating auto emissions policy, since the auto is not a major contributor of these pollutants.

The EPA, the National Institutes of Health, the states, New York City, California Air Pollution Control Districts, industry, the World Health Organization, and other agencies and organizations continue to support the types of toxicologic and epidemiologic studies summarized in the air quality criteria documents. Although they continually add knowledge to the field, these organizations do not provide the information needed to predict benefits from incremental reductions in auto emissions.[11] A project to provide such data would consist of a series of systematic toxicologic studies of people in different health susceptibility categories for various levels of low pollutant concentrations found in ambient air. Another project might monitor the health of people before and after moves to higher or lower pollutant concentration areas. These methods, of course, must confront all the difficulties entailed in measuring the health effects of low levels of pollutants.

CONCLUSION

The approach developed here involves disaggregating benefits (prevented health damage) into smaller, more measureable, categories (e.g., effects on asthmatics); estimating the magnitude of effects on each category for different incremental policy requirements (50, 75, 90 percent emissions reductions); reaggregating to the original categories (health effects of CO, oxidants); and valuing the measured effects from the standpoint of the people affected. It is not known whether the implementation of such an approach before the current national policy was formulated would have resulted in different recommendations by federal officials, scientists, and other parties to the decision. Advice about marginal benefits from auto emissions reduction, especially beyond 75 percent reduction, may yet be needed if adjustments to current policy are contemplated. The tentative conclusion reached here is that benefits from auto emissions reduction rapidly diminish after a 75 percent reduction in CO, HC, and NO_x emissions.

---

[11]Research on damage other than health damage is also needed. The air quality criteria documents contain sections that summarize studies on the effects of pollutants on vegetation and materials. The sections on toxicology deal with animals. Aesthetic reactions to pollutant levels have been gathered by means of surveys. Some of these are presented in the criteria document on photochemical oxidants [U.S. Department of Health, Education, and Welfare 1969a]. The Stanford Research Institute has published large studies on the economic impact of air pollutants on plants [1970].
        Studies of impacts on residential values include those by Anderson and Crocker [1969], Nourse [1967], and Ridker and Henning [1967]. More studies are summarized by Kneese [1967].

References

Andreatch, A.J., Elston, J.C., and Lahey, R., 1971. *The New Jersey repair project, tuneup at idle.* Trenton, New Jersey, June 27, 1971.

Altshuller, A.P., and Bufalini, J.J., 1971. "Photochemical aspects of air pollution: a review. *Environmental Science and Technology* 5:39-64.

Anderson, R., and Crocker, T., 1969. Air pollution and residential property values. Presentation at Econometric Society meeting, New York, 1969.

Appleman, J.M., 1973. A model of auto emissions inspection and maintenance. Unpublished paper prepared for the Automotive Air Pollution Project, Kennedy School of Government, Harvard University, May 1973.

Barrett, L.B., and Waddell, R., 1972. The cost of air pollution damages, a status report. In *Cumulative regulatory effects on the cost of automotive transportation.* Washington, D.C.: Office of Science and Technology.

Barth, D.S., et al. 1970. Federal motor vehicle emission goals for CO, HC, and NO_x based on desired air quality levels. In *Air pollution–1970*, Part 5. U.S. Senate, Committee on Public Works, Subcommittee on Air and Water Pollution, 91st Congress, 2d Session. Washington, D.C.: U.S. Government Printing Office.

Boston Globe, To ease air pollution: tax proposed on Boston parking lots. January 6, 1972, p. 20.

Bruderreck, H., Schneider, W. and Halasz, I., 1964. Quantitative gas chromatographic analysis of hydrocarbons with capillary columns and flame ionization detector. *Analytical Chemistry* 36:461-473.

Center for Environmental Studies, Pennsylvania State University 1971. *Guide to research in air pollution.* Washington, D.C.: U.S. Government Printing Office.

Chandler, A., 1969. *Strategy and structure.* Cambridge, Mass.: MIT Press.

Chase Econometric Associates 1971. *Phase II of the economic impacts of meeting exhaust emission standards, 1971-1980.* Washington, D.C., December 1971.

Chrysler Corporation 1971. Statement by S.L. Terry to the EPA Auto Emissions Standards Hearing, Washington, D.C., May, 1971.

Dewees, D.N., 1971. Automobile air pollution: an economic analysis. Unpublished dissertation and Environmental Systems Program paper, Harvard University.

Environmental Protection Agency 1971a. *Air quality criteria for nitrogen oxides.* Air Pollution Control Office Publication No. AP-84. Washington, D.C.: U.S. Government Printing Office.

Environmental Protection Agency 1971b. *Environmental lead and public health.* Air Pollution Control Office Publication No. AP-90. Washington, D.C.: U.S. Government Printing Office.

Environmental Protection Agency, Office of Air and Water Programs, Mobile Source Pollution Control Program 1972. *Control strategies for in-use vehicles.* Washington, D.C., November 1972.

Esposito, J., 1970. *Vanishing air.* New York: Grossman.

Esso Research and Engineering Company. 1971. Presentation by Neil V. Hakala to the EPA Auto Emissions Standards Hearing, Washington, D.C., May 1971.

Federal Register 35 (November 10, 1970), pp. 17288-17313.
Federal Register 36 (April 30, 1971), p. 8187.
Federal Register 36 (July 2, 1971), pp. 12652-12664.
Federal Register 36 (August 14, 1971), p. 15487.
Federal Register 37 (November 8, 1972), pp. 23778-23779.
Federal Register 37 (November 15, 1972), pp. 24251-24320.

Federal Trade Commission 1968. Staff report on automobile warranties. Unpublished report.

Hocker, Arthur J., 1971. Exhaust emissions from privately owned 1966-1970 California automobiles—a statistical evaluation of surveillance data. Supplement to Progress Report #22, California Air Resources Board, Air Resources Laboratory, April 19, 1971.

Kneese, A., 1967. Economics and the quality of the environment— some empirical experiences. In *Costs of air pollution,* ed. M. Garnsey and J. Hibbs. New York: Praeger.

Kramer, R.I., and Cernansky, N.P., 1970. Motor vehicle emission rates. Internal document, National Air Pollution Control Administration.

Lave, L., and Seskin, E., 1970. Air pollution and human health. *Science* 169 (August 21, 1970): 723-733.

Lave, L., and Seskin, E., 1971. *Does air pollution cause ill health?* Pittsburgh: Carnegie-Mellon.

Learned, E., 1961. General Motors corporation. Harvard Business School Case Study, Cambridge, Mass.

Lees, L., et al. 1972. *Smog—a report to the people.* California Institute of Technology, Environmental Quality Laboratory.

Lowenthal, Mark M., 1971. The industrial conversion of the American automobile industry during World War II. Unpublished paper prepared for the Automotive Air Pollution Project, Kennedy School of Government, Harvard University, December 1971.

Martin, Michael K., 1973. Calculation of aggregate automotive emissions. Unpublished masters thesis, Massachusetts Institute of Technology, January 1973.

McKean, R., 1968. The use of shadow prices. In *Problems in public expenditure analysis,* ed. Samuel B. Chase. Washington, D.C.: Brookings Institution.

National Academy of Sciences 1972. *Semiannual report by the Committee on Motor Vehicle Emissions to the EPA.* Washington, D.C., January 1972.

National Academy of Sciences 1973. *Report by the Committee on Motor Vehicle Emissions.* Washington, D.C., February 1973.

Nelson, D.M., 1946. *Arsenal of democracy.* New York Harcourt, Brace.

New York City Environmental Protection Agency 1972. New York regional air pollution plan, New York City, January 1972.

New York State Department of Environmental Conservation 1972. New York City metropolitan area air quality implementation plan. Albany, New York.

New York Times Encyclopedic Almanac 1970. New York: New York Times.

Northrop Corporation 1971. Mandatory vehicle emission inspection and maintenance. Los Angeles, California.

Nourse, H., 1967. The effect of air pollution on house values. *Land Economics* 43:181-189.

Office of Science and Technology 1970. Report of the Ad Hoc Panel on Unconventional Vehicle Propulsion. Internal document, Washington, D.C., March 1970.

Office of Science and Technology 1972. *Cumulative regulatory effects on the cost of automotive transportation (RECAT).* Washington, D.C., February 1972.

Ridker, R.G., 1967. *Economic costs of air pollution.* New York: Praeger.

Ridker, R.G., and Henning, V., 1967. The determinants of residential property values with special reference to air pollution. *Review of Economics and Statistics* 49:246-257.

Schelling, T.C., 1968. The life you save may be your own. In *Problems in public expenditure analysis,* ed. Samuel B. Chase. Washington, D.C.: Brookings Institution.

Sloan, A., 1964. *My years with General Motors.* Garden City, N.Y.: Doubleday.

Stanford Research Institute (for the Coordinating Research Council) 1970. *Economic impact of air pollutants on plants.* Washington, D.C.: U.S. Department of Commerce, National Technical Information Service.

System Development Corporation (for the Coordinating Research Council) 1971. *A survey of average driving patterns in six urban areas of the United States: summary report.* Washington, D.C.: U.S. Department of Commerce, National Technical Information Service.

TRW Systems Group (for the Coordinating Research Council and the EPA) 1971. The economic effects of mandatory engine maintenance for reducing vehicle exhaust emissions. Redondo Beach, California, August 9, 1971.

U.S. Bureau of the Census 1968. *Statistical abstract of the United States*. Washington D.C.: U.S. Government Printing Office.

United States Code, Volume 42, Section 1857, c-5 (a) (2) (G) (1970).

U.S. Department of Health, Education, and Welfare 1969a. *Air quality criteria for photochemical oxidants*. National Air Pollution Control Administration Publication No. AP-63. Washington, D.C.: U.S. Government Printing Office.

U.S. Department of Health, Education, and Welfare 1969b. *Air quality criteria for sulfur oxides*. National Air Pollution Control Administration Publication No. AP-50. Washington, D.C.: U.S. Government Printing Office.

U.S. Department of Health, Education, and Welfare. 1970a. *Air quality criteria for carbon monoxide*. National Air Pollution Control Administration Publication No. AP-62. Washington, D.C.: U.S. Government Printing Office.

U.S. Department of Health, Education, and Welfare 1970b. *Air quality criteria for hydrocarbons*. National Air Pollution Control Administration Publication No. AP-64. Washington, D.C.: U.S. Government Printing Office.

U.S. Department of Health, Education, and Welfare 1970c. *Air quality criteria for particulate matter*. National Air Pollution Control Administration Publication No. AP-49. Washington, D.C.: U.S. Government Printing Office.

U.S. Department of Health, Education, and Welfare, National Air Pollution Control Administration 1970d. *Control techniques for carbon monoxide, nitrogen oxide, and hydrocarbon emissions from mobile sources*. NAPCA Publication No. AP-66. Washington, D.C.: U.S. Government Printing Office.

U.S. House of Representatives, Subcommittee of the Committee on Appropriations 1971. *Agriculture—environmental and consumer protection appropriations for 1972*, Part 5, 92d Congress, 1st Session. Washington, D.C.: U.S. Government Printing Office.

U.S. Public Health Service 1971a. Acute conditions, U.S., July 1966-June 1967. In *Vital and health statistics*. Washington, D.C.: U.S. Public Health Service.

U.S. Public Health Service 1971b. Chronic conditions, U.S., July 1965-June 1967. *Vital and health statistics*. Washington, D.C.: U.S. Public Health Service.

U.S. Senate, Committee on Commerce and the Subcommittee on Air and Water Pollution of the Committee on Public Works 1967. *Electric vehicles and other alternatives to the internal combustion engine*. Washington, D.C.: U.S. Goverment Printing Office.

U.S. Senate, Committee on Commerce and the Subcommittee on Air and Water Pollution of the Committee on Public Works 1968. *Automobile steam engine and other external combustion engines*. Joint Hearings, 90th Congress, 2d Session. Washington, D.C.: U.S. Government Printing Office.

U.S. Senate, Committee on Public Works, Subcommittee on Air and Water Pollution 1970. *Air pollution 1970*, Part 5, 91st Congress, 2d Session. Washington, D.C.: U.S. Government Printing Office.

White, L., 1971. *The automobile industry since 1945*. Cambridge, Mass.: Harvard University Press.

Index

AAPS (Advanced Automotive Power Systems), 14, 58
Ad Hoc Panel on Unconventional Vehicle Propulsion, 14
Adreath, A., Elston, J. and Leahy, R., 118, 134
aggregation, 71–73
air, ambient, 145, and NO_x, 146
Altshuller, A. and Bufalini, J., 80
AMA (Automobile Manufacturer's Association,), 9, 163; warranty, 120
American Motors, 121
Appleman, J., 134
Army Tank Command, 20, 58
assembly line: testing on, 96–98
Automotive Technical Development Fund, 58, 59
averaging, in testing, 87

balance-of-payments, 40
Barrett, L. and Waddell, R., 176
Barth, Delbert, 13, 163; oxidants, 72
Birmingham, Alabama, 142
Boston: CO, 145, 181
Bruderreck, H., Schneider, W. and Halasz, I., 76

Clean Air Act Amendment, 63; manufacturers, 49
Clean Air Act (CAA): ambiguity, 12; energy use, 24; history, 10; inspection, 34; options, 43; provisions, 29–32; revisions, 56, 57; standards, 173; structure, 4, 5
Cahill, T., 119
California, 29, 116; inspection, 34, 118; oxidants, 161
California Institute of Technology, 9, 182
CAMP (Continuous Air Monitoring Project) 146, 167, 168, 179; and CO, 181
Chattanooga: NO_2, 159, 160, 172
CHESS (Community Health Effects Surveillance Studies), 203
Chevrolet: design, 123
Chicago, 13; CO, 145, 164–166
Chrysler Corp., 11, 54; warranty, 121
cigarette smoking, 157
Clean Car Race, 11
competition, 5
Congress, 50, 67; and politics, 55; and standards, 86; and support, 14
correlation: in Weinstein and Clark, 90
costs: and ICE, 18, 19; in-use testing, 115, 116; and options, 35, 37, 38; social of pollutants, 185–203
crisis of the commons: as a definition, 3
CVS (constant volume sampling), 66, 69, 96

David, Edward, 11
day of restricted activity: definition in Ahern, 190
Delaware, 116
Denver: oxidants, 168, 179
Dept. of Transportation, 60
design: and warranty, 122
deterioration: and durability, 95; and emissions, 83
Donora, 142
Durant, William, 50

Earth Day, 11
EDRA, 189–205
EGR (exhaust gas recirculation), 15; in Wankel, 20
electric drive, 23, 24
emissions: and adjustment, 150; definition in Ahern, 142; definition in Weinstein and Clark, 72; difficulty in control, 15,

211

16; forecast, 30–34; and health, 187–204; in ICE and alternatives, 20–25; inspection, 96–98, 124–137; modes of measurement, 74–83; NO_x, 15; options, 32, 33; rates and tests, 2; and social costs, 178
Environmental Protection Agency (EPA), 11; mandate, 12; framework, 64–66; Air Pollution Control Office, 177
Energy Research and Development Administration, 59
enforcement,: and instruments, 68; lack of, 130–133; in Weinstein and Clark, 73
Environmental Quality Laboratory, 182
Esposito, John, 11
exposure: and emissions, 147; to pollutants, 184
"external effects," as a definition in Jacoby, and Steinbruner, 3

FID (flame ionization detection), 76
Ford Motor Co., 20, 52, 54, 58, 120; warranty, 121
FTC (Federal Trade Commission): warranty, 121
fuel: consumption and cost, 19; economy, 15

gas turbine, 22, 23, 52
General Motors, 50, 52, 54, 121
GNP (Gross National Product), and ICE, 1; and pollutants costs, 176
government: and implementation, 5

Hacker, A., 126
health: and pollutants, 152–162; and susceptibility, 187–204
HEW (Dept. of Health, Education and Welfare), 9
Honda, 21
House Ways and Means Committee, 59

ICE (internal combustion machine): alternatives, 20–24; gaseous fuels, 24, 25; and instability, 39; and GNP, 1
Idle Test, 67, 69, 96, 97
implementation: and gas turbines, 23; scope of, 3; strategy, 54–61
innovation, 51
inspection: on assembly-line, 96–98; and implementation, 100; on the road, 125; and resistance, 117

Jackson, Henry, 11
Justice Dept., 11, 60

Kettering, Charles, 51
Key Mode Test, 68–70

Kramer, R. and Cernansky, N., 163

Lave, L. and Seskin, E., 175
Learned, E., 122
Lewenthal, M., 53
London, 142
Los Angeles, 13; and cars, 142; and CO, 145, 154–158, 164; and NO_2, 173; and O_3, 146; and oxidants, 168, 179, 184, 196
Los Angeles Air Pollution Control District, 9
Los Angeles County Board of Supervisors, 10

McKean, R., 176, 201
maintenance, 17, 18; and car owners, 66–68; and control system, 124; and deterioration, 90–92; enforced, 34–37, 133; and inspection, 30; and options, 43
manufacturer: and recall, 120–123
Mass. Bureau of Air Quality Control, 119
measurement: and instruments for, 76; modes of, 74–83; problems of, 69–71; samples, 94
models, mathematical, 105–111; in Ahern, of exhaust, 139, 141
Muskie, Edmund, 11, 163; deadline, 14

Nader, Ralph, 10, 11
NAS (National Academy of Science): and deadline, 12; and Honda, 21; and implementation, 25, 26; inspection, 18; testing report, 64; test variation, 117
National Center for Health Statistics, 188
National Highway Safety Act, 120
National Science Foundation, 58
NBC News, 1
Nelson, Donald M., 53
Nelson, Gaylord, 11
New Jersey, 116; inspection, 34, 118, 134
New Jersey Pollution Control Act, 119
New York: and cars, 142; CO, 145
New York Environmental Protection Agency, 116, 119
New York State Department of Environmental Conservation, 119
New York Times, 11
Nixon, Richard, 11, 59; message to Congress, 14
Northrup Corp., 116

Office of Management and Budget, 14, 40, 59
oil: consumption, 40
options: ALT, 27; CO, 32, 33; and EST (established policy), 27; EST and projections, 41; EST/ALT, 27, 39, 43; EST/ENF (established policy and enforcement program), 27, 43; EST/

ENF and maintenance, 35; implementation, 27–29; REL, 27, 36–38, 43; REL/ALT, 27, 39, 43; REL/ALT, EST/ALT, 49
OST (Office of Science and Technology), 14, 34
Otto cycle ICE, 52
outcome, calculation of, 5, 6
oxidants, 166–174; and CAMP, 184; and monitor system, 180

Pasadena: reductions, 179
performance, 101–104
Philadelphia, 33; oxidants, 168
politics: and extensions, 25, 26; growing consciousness, 11; inspection, 118; policy, 196–204; and standards, 38; and testing, 88; and implementation, 58–61
pollutants: and –cost, 176; definition in Ahern, 141–143; and health, 152; in ICE and alternatives, 20–25; measurements, 74–83; and options, 32, 33; reduction, 183; and residential values, 205; social cost, 178–187; and susceptibility, 189–206
PROCO (Programmed Combustion program), 58
production: manufacturer and testing, 83
Public Health Service, 10, 188

quenching, definition of, 20

Rankine Cycle, 21, 53; and costs, 39; HC, 22; and NO_x, 22
reallocations: in Jacoby and Steinbruner, 57–61
recall, 99
regionalization, 34
respiratory system: and exposure, 148–150
Richardson, Elliot, 11, 163
Ridker, R., 175
Ruckelshaus, William, 11, 203; extensions, 25

Sacramento: reductions, 179
San Diego: reductions, 179
St. Louis: and cars, 142
sampling, in testing, 82
Seven Mode Hot Cycle Test, 67
"Short" Cycle Test, 67, 69
stability, 16, 18; definition, 30; and maintenance, 35; and technology, 44–47
standards: 1975–76, 2; relaxation of, 37
State Motor Vehicle Control Board, 10
Stratified Charge ICE, 20

tampering: emissions, 126
technology, 29, 38–40; and health effects, 186–203; incentives, 57, 58; options, 45–47
testing, 29; averaging, 87; costs and implementation, 104; enforcement, 114; in-use, 114–137; kinds of, 67; on assembly lines, 96–98; pollutants and health, 155–162; procedures, 16–18; prototypes, 92–95; samples, 74–83
toxology: and pollutants, 156–162
trade-offs: in implementation, 4
Treasury Dept., 59

U.S. Court of Appeals, 26
U.S. Senate Committee on Public Works, 120

Vanishing Air, 11

Wankel, 19, 20, 21, 52; and HC, 20; and NO_x, 26
warranty, 68; and maintenance, 100; and manufacturer, 120; and performance, 114
Washington, D.C., 116; and CO, 145; and oxidants, 168
White, L., 54
World War II, 53

About the Authors

      Henry D. Jacoby received his Ph.D. in economics from Harvard University in 1967, was Associate Professor of Political Economy at the John F. Kennedy School of Government, Harvard, and is presently Professor of Management at M.I.T. He is co-author of a new book entitled MODELS FOR MANAGING REGIONAL WATER QUALITY.

      John D. Steinbruner received his Ph.D. in political science from M.I.T. in 1968 and is now Associate Professor of Public Policy at the Kennedy School. His book, tentatively entitled THE CYBERNETIC THEORY OF DECISION, will be published this winter.

      Professors Jacoby and Steinbruner were Co-Directors of the Automotive Air Pollution Project, a three-year joint effort of the Environmental Systems Program and the Institute of Politics, Harvard, sponsored by grants from the National Science Foundation and the Ford Foundation. William R. Ahern, Jr., Jack M. Appleman, Ian D. Clark, and Milton C. Weinstein were all students in the Public Policy Program of the Kennedy School at the time of their collaboration with Professors Jacoby and Steinbruner on this project. William Ahern is now a policy analyst with the RAND Corporation in Santa Monica, California (his book, OIL AND THE OUTER COASTAL SHELF: THE GEORGES BANK CASE, is also being published by Ballinger); Jack Appleman is a teaching fellow at the Kennedy School and a research fellow at the Woods Hole Oceanographic Institute in Massachusetts; Ian Clark is an executive assistant to the Minister of State for Urban Affairs, Ottawa; and Milton Weinstein is Assistant Professor of Public Policy at the Kennedy School.